Cambridge Studies in Management

14

Managing innovation

Cambridge Studies in Management

Formerly Management and Industrial Relations series

Editors

WILLIAM BROWN, *University of Cambridge*
ANTHONY HOPWOOD, *London School of Economics*
and PAUL WILLMAN, *London Business School*

The series focuses on the human and organisational aspects of management. It covers the areas of organisation theory and behaviour, strategy and business policy, the organisational and social aspects of accounting, personnel and human resource management, industrial relations and industrial sociology.

The series aims for high standards of scholarship and seeks to publish the best among original theoretical and empirical research; innovative contributions to advancing understanding in the area; books which synthesize and/or review the best of current research, and aim to make the work published in specialist journals more widely accessible; and texts for upper-level undergraduates, for graduates and for vocational courses such as MBA programmes. Edited collections may be accepted where they maintain a high and consistent standard and are on a coherent, clearly-defined, and relevant theme.

The books are intended for an international audience among specialists in universities and business schools, undergraduate, graduate and MBA students, and also for a wider readership among business practitioners and trade unionists.

Other books in the series:

1 John Child and Bruce Partridge, *Lost managers: supervisors in industry and society*
2 Brian Chiplin and Peter Sloane, *Tackling discrimination in the workplace: an analysis of sex discrimination in Britain*
3 Geoffrey Whittington, *Inflation accounting: an introduction to the debate*
4 Keith Thurley and Stephen Wood (eds.), *Industrial relations and management strategy*
5 Larry James, *Power in a trade union: the role of the district committee in the AUEW*
6 Stephen T. Parkinson, *New product development in engineering: a comparison of the British and West German machine tool industries*
7 David Tweedale and Geoffrey Whittington, *The debate on inflation accounting*
8 Paul Willman and Graham Winch, *Innovation and management control: labour relations at BL Cars*
9 Lex Donaldson, *In defence of organisation theory: a reply to the crisis*
10 David Cooper and Trevor Hopper (eds.), *Debating coal closures: economic calculation in the coal dispute 1984–5*
11 Jon Clark, Ian McLoughlin, Howard Rose and Robin King, *The process of technological change: new technology and social choice in the workplace*
12 Sandra Dawson, Paul Willman, Martin Bamford and Alan Clinton, *Safety at work: the limits of self-regulation*
13 Keith Bradley and Aaron Nejad, *Managing owners: the National Freight Consortium in perspective*

Managing innovation: A study of British and Japanese factories

D.H. WHITTAKER

Lecturer in Japanese Studies, Cambridge University

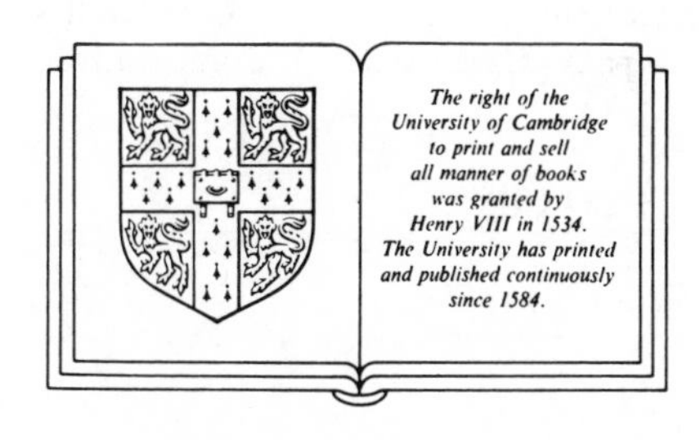

CAMBRIDGE UNIVERSITY PRESS

Cambridge
New York Port Chester
Melbourne Sydney

Published by the Press Syndicate of the University of Cambridge
The Pitt Building, Trumpington Street, Cambridge CB2 1RP
40 West 20th Street, New York, NY 10011, USA
10 Stamford Road, Oakleigh, Melbourne 3166, Australia

First published 1990

Printed in Great Britain at the University Press, Cambridge

British Library cataloguing in publication data

Whittaker, D.H.
Managing innovation: a study of British and Japanese factories.
1. Great Britain. Industries. Technological innovation. Management.
2. Japan. Industries. Technological innovation. Management aspects.
I. Title
658.4'063'0941

Library of Congress cataloguing in publication data

Whittaker, D.H. (D. Hugh)
Managing innovation: a study of British and Japanese factories / D.H. Whittaker.
p. cm.
Bibliography.
Includes index.
ISBN 0–521–38055–3
1. Technological innovations – Great Britain – Management – Case studies. 2. Technological innovations – Japan – Management – Case studies. 3. Industrial management – Great Britain – Case studies. 4. Industrial management – Japan – Case studies. 5. Comparative management. I. Title.
HD45.W44 1990
658.5'14'0941 – dc20 89–33234 CIP

ISBN 0 521 38055 3

SE

Contents

List of figures *page* vii
List of tables viii
Foreword xi
Acknowledgements xvi

1 British factories, Japanese factories and the new technology debate 1
1.1 Introduction 1
1.2 Employment relations 5
1.3 Employment relations, computerized machine tools (CNC) and skills: four hypotheses 10
1.4 Eighteen factories 17

2 The wider context 20
2.1 'Lifetime employment' and 'market contractualism': a macro view 20
2.2 Shaping factors 29
2.3 Recent changes 48

3 Employment relations 1 53
3.1 Employment 54
3.2 Payment systems 66

4 Employment relations 2 80
4.1 Industrial relations 80
4.2 OER, MER and the 18 factories 94

5 Innovation 99
5.1 Introducing CNC 99
5.2 'Soft' process innovation 108
5.3 CNC – isolated innovation, or linked? 111

6 Training 114
6.1 General training 115
6.2 Training for CNC 128
6.3 Training, OER and MER 133

7 Division of labour 138
7.1 Operators' tasks 139
7.2 Programming 144
7.3 Task ranges, OER and MER 151

8 CNC use and skills 156
8.1 Craft versus technical approach 156
8.2 CNC, skills and deskilling 158
8.3 Employment relations and new technology 161

Appendix 1 Glossary of technical terms 164
Appendix 2 The 18 factories: a brief introduction 165

Notes 172
Bibliography 187
Index 201

Figures

1.1 Career training for CNC operators in a Japanese factory 12
2.1 Age-earnings profiles in Britain and Japan 24
2.2 Stages of corporate control 39
3.1 Average lengths of employment by age, J45 and B71 59
3.2 Pay system concept, J45 67
3.3 'Standard' promotion, J45 69
3.4 Age-earnings of union members, J45 71
3.5 Age-earnings profiles at B71 and J45 75
4.1 OER–MER continuum 98
7.1 CNC operator tasks 140

Tables

1.1 Organization-oriented employment relations (OER) and market-oriented employment relations (MER) 7
1.2 The 18 factories 18
1.3 The 18 factories; batch sizes 19
1.4 Changes in employment, 1977–87 19
2.1 'Fixed employment' in Japan and Britain 21
2.2 Employment length by occupational category (Britain) 22
2.3 'Fixed employment': manual and non-manual workers 23
2.4 Membership by type of union (Japan) 25
2.5 Employment by company size (Japan) 36
2.6 Labour costs by company size (Japan) 37
2.7 Pay differentials by size of establishment 38
2.8 Share ownership in Japan and Britain 40
3.1 Recruitment, f1986 55
3.2 Average ages and lengths of employment (manual workers) 58
3.3 Backgrounds and tertiary education levels of selected managers 60
3.4 Ex-shop-floor workers in various departments 61
3.5 Wage components, J45 70
3.6 Wages of direct operators, B71 73
4.1 Formal industrial relations organization 81
4.2 Negotiating/consultation bodies in the larger British factories 84
4.3 OER–MER scores 97
5.1 Introduction of CNC 100
6.1 Time-served CNC operators (British factories) 118
6.2 Prior experience on manual machines 123
6.3 Ages of CNC operators 126
6.4 Ages of CNC and manual operators 127
6.5 CNC-specific training 129
6.6 Training scores 134

6.7 Rank orders: training, OER–MER, size, batch size 135
6.8 Training/OER–MER, size, batch size (Spearman rank correlations) 136
7.1 Programmers 149
7.2 CNC operator task range 152
7.3 Task range, OER–MER, size, batch size, training (Spearman rank correlations) 154
8.1 Differences in CNC use and attitudes to CNC: Japan and Britain 157

Foreword

Japan has been getting closer. Soon there will be over a hundred factories in Britain managed by Japanese firms. A few years ago the spread of 'quality circles' acquired something of the character of a missionary movement. Today it is 'just-in-time' manufacture which is offered as the panacea. Books and articles on Japan abound. Most people in industry, in the universities and in journalism have got some idea that the Japanese organize their affairs rather differently from the way we do, even if they would have to resort to hazy notions of loyalty, paternalism and lifetime employment if they were asked to describe in detail how they differ.

So what need for another book on Japanese industry, some might say, after a superficial glance at Hugh Whittaker's book. But, in fact, this is not just another book on Japanese industry. First, the book is even-handedly as much about Britain as it is about Japan; a careful point-by-point comparison not only of the nine British with the nine Japanese factories in which he did his detailed field-work, but also of the national structures in which those eighteen factories were embedded. And it is a comparison which illuminates. We learn more about Japan from its being juxtaposed, point by point, with Britain: and vice versa.

And the second point is that Whittaker does not accept that *national* differences are the only kinds of differences to look for or the only kind that can be instructive. His awareness of the variety of patterns *within* Britain and *within* Japan leads him to think in terms of 'types of employment relations' – one type predominating in each country, to be sure, but by no means in perfect correlation. And he is able, thereby, to link up with an important tradition of thinking about how large-scale work organizations can be run.

Ever since Robert Owen, thoughtful capitalists and managers have been casting about for some way round the awkward fact that efficient production requires cooperative relationships within hierarchies – between those who do the planning and the giving of orders, and those who take the orders and yet at the same time they have to enter into a contractual wage-effort bargaining relationship which by its very nature is adversarial. Two kinds of solution have been offered to the problem, and Robert Owen experimented with both. One kind may be called capitalism-changing solution: the other within-capitalism solution. The first consists of various proposals for the radical restructuring of the ownership function – cooperatives, employee ownership trusts, profit-share schemes – which remove, or substantially modify, the adversarial element in wage determination. The other kind, staying within the conventional capitalist framework with its trade-off between wages and dividends which at least in the short term is zero-sum, has involved styles of management – human relations skills, in one much-favoured formulation. After all, some would argue, in the long term the trade-off between wages and returns to capital need *not* be zero-sum; there can be significant joint gains if 'the two sides' of industry cooperate with each other. And in any case, all human relationships, from marriage or parenthood to customer/store clerk contacts, contain both cooperative and zero-sum adversarial elements. It is always possible, by taking thought, to accentuate the cooperative and play down the adversarial.

So, for instance, MacGregor, with his X-theory and Y-theory, suggested that if you treat people as if you expected them to be ready to put effort into doing a good job and making the firm prosper – refrained from suspiciously checking up on them, for instance – they would actually work in a way conforming to those expectations. Alan Fox took the discussion further in his distinction between a 'high-trust firm' in which workers can be expected to work 'responsibly' because managers accept reciprocal responsibilities (especially the responsibility to be honest) towards the workers, and a 'low-trust firm' in which any would-be Y-theory manager would be likely to suffer grievous disappointment and revert to suspicious, 'give 'em an inch and they'll take an ell', X-theory assumptions.

Here, Whittaker has used his empirical observation of the normal, standard practice in Britain and Japan, and combining it with these theoretical ideas, to spell out in some detail the sort of *organizational* characteristics, the 'type of employment relations', which one would expect to lead to 'high trust' cooperativeness – 'organization-oriented employment relations' – and the alternative type – 'market-oriented employment relations' – which one would expect to lead to an accentuation of adversarial wariness. The former is characterised by long-term job security as opposed to short-range hire and fire; rate-for-the-person incremental-scale, as opposed to rate-for-the-job, wage structures; relatively predictable career promotion patterns as opposed to wholly discretionary promotion patterns and uncertain prospects; union structures that reinforce a sense of organization membership as opposed to occupational membership.

And, lo and behold, when he scores his firms on a variety of these characteristics, the correlation between type and country is not perfect. There is some overlap. There are two Japanese firms which are more 'market-oriented' than at least five of the British ones.

It is worth noting, apropos the distinction made earlier between capitalism-modifying solutions and within-capitalism solutions to the cooperation/adversarial opposition problem, that the two Japanese firms with 'market-oriented employment relations' are small, individually-owned, 'pure capitalist' firms. Larger Japanese firms, as Whittaker points out in his second chapter, do significantly constrain exercise of the ownership function. There are many elements in this; high debt–equity ratios; interlocking patterns of shareholdings by a firm's bankers, suppliers, insurers, distributors; minimal year-to-year variations in dividend levels; above all an effective taboo on hostile take-overs, reflecting a dominant social perception of the firm as a community of people rather than as a piece of shareholders' property. The shareholder is not as sovereign in practice as Japanese company law says he is – or as, *both* in law and practice, he is in Britain.

So the fact that the two 'market-oriented' Japanese firms are firms to which all these constraints on the ownership function do not apply may be significant. It may mean that effective, wholly 'within capitalism' – at least 'within pure capitalism' – solutions to the problem do not exist. Some modification of the absoluteness of capital-ownership rights may be a precondition for any successful attempt to accentuate the cooperative, and minimize the adversarial elements in the wage relationship.

But Whittaker takes the story very much further than that. He asks about the efficiency aspects of these patterns of employment relations, which he studies through careful examination of one particular pattern of change – a change which all his 18 factories had experienced – namely, the introduction of new machine tools controlled, not by men's eyes and hands like traditional machine tools, but by computers.

His findings are interesting, and sometimes counter-intuitive. Japanese firms give workers put on to the new CNC machines not more training than do the British firms, but less. But they do seem to expect faster and more extensive learning on the job, and more learning from books. And they have a more broadly defined range of tasks to perform. Factory size seems a potent determinant of training patterns in both countries; the difference between standardized large-batch work and specialized small-batch work less so.

In exemplary, textbook fashion, Whittaker framed his data collection around a set of hypotheses concerning training and the division of labour. In the end, as will occasion no surprise to anyone except the people who write the textbooks, his most interesting observations have to do with something other than the testing of his hypotheses – namely the difference, of which he gradually became aware, between what he calls the 'technical' approach which dominated in Japan, and the 'craft' approach which dominated in Britain. ('Technician approach' and 'craftsman approach' might be alternative names.) The Japanese are ready to turn their craftsmen into technicians – to embrace CNC machines as a means of replacing uncertain audio/visual/manual skills which depend on 'knack', with precise intellectual skills for solving problems with right-or-wrong answers. British workers and managers, by contrast, were reluctant to give up the idea that the CNC machines should only be run by people who had a feel for metals and could tell from the sound of the tool biting into the work-piece whether the speeds and feeds were right or not.

And in the end this distinction has to be expressed as a Britain/Japan difference, not a difference between organization-oriented and market-oriented employment relations, because, as Whittaker shows, it is a difference determined by *society-wide* and not *organization-specific* variables. 'Media-climates' – what the newspaper and trade magazines say about CNC machines and what counts as success in using them – are national. So are educational systems, and Japanese schools produce much higher levels of numeracy and more people with the intellectual self-confidence to learn from instruction manuals, than the British. Faster rates of growth have made the Japanese more generally enthusiastic about technical change, more willing to write off the old and cheer in the new. (Or is it that more gung-ho attitudes towards change have produced faster rates of growth?) A society in which organization-oriented employment patterns are the *norm* does not have the craft unions of a market-oriented society like Britain, hence does not have the apprenticeship patterns that go with them – nor the mystique of craft, nor the hard status privileges attached to skill labels which their owners naturally resist seeing eroded.

All of which would seem to point to the moral that, if you want to make industry more efficient, it is not enough to rely chiefly on exhorting *employers* to mend their ways, while taking care of the societal variables only by squashing the unions and creating City Technology Colleges. Just how much the characteristics of society as a whole constrain attempts to alter the organization and ethos of particular firms within it is shown in Whittaker's analysis of the extent to which British firms are edging along his continuum towards more organization-oriented patterns. One firm's attempts to change its job-grading system (for greater 'flexibility'), and to 'harmonize' (to give the same conditions of employment to both manual and non-manual workers) may have effects on work practices and attitudes (in the cooperativeness/adversarial antagonism dimension, that is to say), but those effects will be seriously limited by the fact that the changes have to work through and against the norms and interest-constellations of the society as a whole.

It is its sensitive handling of the complexity of all these dimensions, as well as its boldness of scale (detailed case studies of one or two factories are usually enough for most researchers), which make Whittaker's book such a valuable contribution. And it is the graphic detail of shop-floor life and work which make it such an interesting book to read. Those who want to know how to run factories efficiently, those who want to know whether Britain is likely to catch up with Japan, those who want to know whether new technology is the means by which capitalists exercise greater control over and extract greater surplus value from their workers, those who want to know how to promote industrial training, and those who just want to read a good piece of sociology, will all get something out of this book.

Cambridge, Massachusetts RONALD DORE

Acknowledgements

I would first like to thank all the people in both the British factories and the Japanese factories I visited for graciously receiving me and patiently answering my many questions. Unfortunately, they must remain nameless, but I have tried to respect and record their experiences and ideas rather than treating them as statistical objects.

I would like to thank Ronald Dore, who gave me invaluable advice and encouragement from beginning to end; also Roderick Martin for his help and many useful comments; Sandra Dawson and other members of the DSES (now Management School) faculty at Imperial College; and William Brown of Cambridge University.

My thanks in Japan go especially to Takeshi Inagami for his help and comments; also to Yasuo Kuwahara, and Osamu Hirota and the staff and researchers of the Japan Institute of Labour; Naoyuki Kameyama of the NIEVR, S. Yahata, O. Yokokawa, T. Koseki, K. Mori and K. Watanabe.

Part of the research in Japan was funded by the Central Research Fund of London University. A fellowship from the Program on US–Japan Relations, Harvard University made the writing up possible, and I would like to express my appreciation to Susan Pharr, Ezra Vogel, Mitchell Sedgwick and the staff and associates of the Program.

My thanks to CUP and especially Patrick McCartan and Trudi Tate for making publication of this book possible, and finally to Toshie, also, for all her patience.

1

British factories, Japanese factories and the new technology debate

1.1 Introduction

This book sets out to answer a limited number of questions but covers a wide range of enquiry, from personnel practices and industrial relations to factory automation, training and work organization. It aims to provide a perspective on how process innovation is approached and new technology used in British and Japanese factories, and how these are related to employment relations. The basic questions asked are these: Why should the same technology introduced into manufacturing organizations invite friction in some cases and cooperation in others, or lead to 'deskilling' in some cases and 'reskilling' in others? What practices, attitudes and institutions lie behind these differences?

The book focusses on a particular application of microelectronics technology – computerized machine tools. Machine tools cut, drill, bore, grind or shape metal with various tools. Conventionally their movements are controlled by levers and handles manipulated by an operator, but numerical control (NC) machine tools are manipulated by a predetermined code or programme initiated via an electronic control system. Computer numerical control (CNC), as the name suggests, adds to this a reprogrammable computerized controller.[1] If machine tools, which cut, grind or shape metal are the 'guts' of modern industry, as Noble (1984) says, NC and CNC have been the 'guts' of much of the debate and occasional research into the usage of new technology.

Much of this was sparked by Braverman, who, in an influential book, *Labor and Monopoly Capital* (1974), argued that new technology in general and numerical control in particular allows capitalists to gain control over the work process from recalcitrant craftsmen and 'deskill' them and their jobs. It does this by separating the conceptual part of the work, which becomes programming, from execution, which becomes machine minding. Tapes can be prepared in an office by workers amenable to managerial control, allowing managers to replace the craftsmen on the shop-floor with unskilled workers and more machines. This possibility is seized upon by capitalists and their agents with the 'inevitability that devastates with the force of a natural calamity' (Braverman, 1974, 194) as it removes the final obstacle to the capitalists' quest for control and the freedom to pursue surplus value in the face of mounting competition.

Other writers have pointed out that there are alternative ways to extract surplus value or pursue profits, and that new technology such as computerized machine tools may be used for these ends. Attempting to reduce delivery times to gain new orders is one example. While increased control and 'deskilling' may indeed represent one management strategy, this may reach the point of diminishing returns, for labour can never be completely commodified and capitalists need not only to control workers, but need their consent as well (Littler and Salaman, 1984). This is especially so where flexibility in work practices is needed to utilize new technology most efficiently. Different management strategies, based on human relations or neo-human relations theories seeking to enlist worker cooperation, may lead to 'reskilling' and job enlargement. Operators may be given the job of programming, which has become more feasible with recent CNC technology.

Whether managers pursue coherent 'strategies' or not is of course open to question. Even if they do, strategies are shaped by existing structures (Batstone *et al.*, 1987), and one strategy might be pursued in industrial relations and another in work processes (Rose and Jones, 1985). One further perspective is that computerized machine tool use is not the result of capitalists' preoccupation with deskilling or managerial strategies at all, but of such factors as factory size and the size of the batches that the factory turns out. Large factories and large batches lead to a bureaucratic division of labour and a polarization of skills. Beyond these, differences from one factory to another are related mainly to production organization and approaches to training (Sorge *et al.*, 1983).

To shed light on this debate, this book looks at computerized machine tool use in Japanese factories, famous for their cooperative industrial relations, and in British factories, famous for a more abrasive style of industrial relations. Industrial relations do not simply reflect managerial strategies; in Japan they are said to be upheld by the 'three pillars' of lifetime employment, *nenko* (seniority plus merit) wages and promotion, and enterprise unionism. The linkages between these often institutionalized aspects of employment relations (employment, payment and industrial relations systems) and computerized machine tool use are therefore explored.

Of course not all Japanese factories are characterized by one type of employment relation and all British factories by another. Two ideal types are constructed; organization-oriented employment relations (OER) and market-oriented employment relations (MER). The terms and basic concepts originate from Dore (1973), and are in effect a formalized and extreme epitomization of what are often spoken of as typical Japanese and British employment relations respectively. OER and MER are poles of a continuum along which the employment relations of different factories may be located. These locations may be compared with certain aspects of CNC use, particularly the training of operators and the organization of operators' tasks, to find out whether or not there are any systematic differences related to employment relations.

This approach should prove more sensitive to both cross-national as well as intra-national differences than blanket typologies like that of Littler (1982), which describes Taylorism as the predominant management strategy in British factories, characterized by a dynamic of deskilling and task control, as opposed to *shudanshugi* (groupism)[2] in Japanese factories, which leads to the development of generalized, semi-skilled workers. (This approach has the additional weakness of placing too much emphasis on management strategies.) First, the continuum itself is more sensitive than are discrete typologies, and secondly, it does not assume that all British companies are located at one end and all Japanese companies at the other.

Four hypotheses are given; that the closer a factory to the organization-oriented (OER) pole of employment relations, the more training is given to CNC operators; that the closer a factory to the OER pole, the wider their task range; that the influence of factory size and batch size is mediated by employment relations; and that the closer a factory to the OER pole, the higher the skill levels of CNC operators.[3]

In the course of testing these hypotheses a number of related, topical issues are discussed. Many of these relate to innovation, both technological and otherwise. British (and American) managers are often criticized for their shortcomings regarding innovation – their apparent failure to take it seriously or reflect it in integrated, long-term strategies. This results in wasted potential, unnecessary friction and sometimes outright failure.[4]

The implicit or explicit exemplary approach held up is frequently that of Japanese firms. Not only are Japanese companies good at commercializing new technologies, as can be seen with computerized machine tools, but they seem to be extremely good at using them, too. Japanese managers are supposedly convinced of the strategic value of new technologies, are able to take a long-term view in corporate planning free from the tyranny of accountants, invest in training their 'human resources' rather than treating them as costs, and so on.

Not only are Japanese managers credited with having a more 'holistic' view of innovation – linking separate microelectronics applications to achieve systems gains (Kaplinsky, 1984) and linking the introduction of new technology to non-technical factors at the planning stage – but they seem to have a more holistic view of the innovation process itself, which they have acquired in the course of post-war reconstruction, in their drive to catch up with the west, and in surviving intense domestic competition. The objectives of introducing new technology – reducing operating costs, improving efficiency, increasing flexibility, raising the consistency and quality of products and improving control over operational processes according to Child (1984) – coincide with those of another form of process innovation developed in Japan and seen by some as being as revolutionary as Taylor's scientific management; just-in-time (JIT) production. The refinement of just-in-time, and one might add quality control activities, often go hand in hand with the introduction of new technology according to Abegglen and Stalk (1985).

These developments are aided by flexible employment practices which stress affiliation to the company rather than to a particular job, and payment and industrial relations practices which also tend to promote the flexibility that microelectronics, blending as it does traditional organizational and job boundaries, is seen to require.[5] Job rotation and career ladders translate potential for flexibility into actuality. These practices are themselves innovations, starting before World War II but taking their present shape in the post-war period.

Some of the accounts of the Japanese 'model' are probably more prescriptive than descriptive, a projection of what would cure our industrial woes independent of their existence – or otherwise – in Japan. Furthermore, what may be true for Toyota or Nissan, Fanuc or Yamazaki, may not hold for other Japanese firms. A careful study of employment relations and computerized machine tool use in a variety of British and Japanese factories should aid our understanding of the 'innovative firm', the 'flexible firm', 'harmonization' and indeed 'Japanization', which provokes sharp reactions but is increasingly talked about.[6]

Nine factories in both countries, matched as closely as possible by size, technology and product, have been selected. While no claim to complete representativeness can be made, they provide a broad picture of the respective mechanical engineering industries which does not focus solely on technology leaders and large, famous firms, as is the case in so many studies of Japanese industry.

The findings point to significant differences both within and between the countries, some of which are surprising in view of conventional wisdom or popular accounts. British operators by and large were given *more* training for CNC than their Japanese counterparts, although their task ranges were not as extensive. This is linked to a different approach to computerized machine tools in both countries which has significant implications for innovation; a 'technical' approach in many of the Japanese factories, and a 'craft' approach in many of the British ones. In the former, for example, unmanned operation was an attraction of CNC, while in the latter a skilled craftsman had to be by the machine to get the most out of it. These approaches were influenced by employment relations, but in a more subtle way than at first suggested.

In the remainder of this chapter the reasoning behind the hypotheses mentioned above and the concepts they draw upon will be outlined, and a brief introduction to the 18 factories given. Chapter 2 provides an overview of the context of employment relations in Britain and Japan, which will help to put the individual factories in perspective. Chapters 3 and 4 describe employment relations in the 18 factories, while chapter 5 describes the introduction of CNC and other aspects of process innovation. Chapter 6 relates the training of workers operating the CNC machines, and chapter 7 the division of tasks around them. The various strands of these chapters are brought together and summarized in chapter 8, and the implications discussed. Hopefully the book will provide a modest contribution to the issues raised above, and to cross-national industrial research and understanding in general.

1.2 Employment relations

What are employment relations?

If the Japanese take a more holistic approach to innovation and corporate activities, as some have suggested, perhaps they do to industrial relations as well. The 'three pillars' or 'three sacred treasures' of lifetime employment, *nenko* (seniority plus merit) wages and promotion, and enterprise unions are the eternal theme of industrial relations discussions in Japan, according to an Economic Planning Agency (1986) book.[7] It is rare to include all of this subject matter – employment practices, for example, and certain aspects of wages and promotion – in discussions of industrial relations in Britain, despite talk about industrial relations 'systems'. In Japan, however, the three are considered to be intimately connected, and all must be considered when speaking of an industrial relations system.

The argument runs something like this: long-term employment, desirable from the company point of view to train and keep human resources, makes it possible and desirable – and is facilitated by – the rewarding of employees over a long time period, hence *nenko* wages and promotion. Workers within such companies find that their interests match closely those of others in the same company, hence they tend to organize within the organization itself – enterprise unions – which in turn promotes internal careers, and so on. The respective pillars are an integral part of the 'system'.

This 'system' I have called 'employment relations'. It includes employment, payment/reward and industrial relations. A further aspect which is seldom discussed under the rubric of industrial relations is ownership. Ownership may be considered an integral part of employment relations, or an (external) influence on them. Here the latter approach is taken, but this is one area in which more work needs to be done by industrial relations and related specialists.

Organization orientation and market orientation

One cannot assume that the three pillars are equally characteristic of all Japanese companies, or that individualistic, job-based contractualism based on external labour markets, which 'Japanese-style management' and the three pillars are often contrasted with, are equally characteristic of all western or British companies. Here I will propose two ideal, polar types of employment relations, the continuum between which different firms may be located; *organization-oriented employment relations* (OER) and *market-oriented employment relations* (MER). As with the three pillars, there are pressures towards inner consistency in the various sub-dimensions of these which would tend to locate companies towards these respective poles, but there are also other forces – economic and social – which act on the various sub-dimensions differentially and in divergent directions. Moreover, different groups of workers within the same company may be treated differently, such as full timers and part timers, or manual workers and non-manual workers. The focus in this book is on manual workers and the shop-floor, where interfirm differences in employment relations are likely to be most obvious. The concepts are summarized in table 1.1.

In OER employment, relative to that of MER, there would be:

more rigorous selection of employees because of the expectation of long-term employment;

more induction training for the same reason;

greater discrimination against mid-term entry and a younger average entry age;

the employment of workers for a career within the company not tied to a specific job.

In OER payment, relative to that of MER, there would be:

limited reference to market rates of pay, and fixed, organization-wide rules regarding pay raises and relativities;

Table 1.1 *Organization-oriented employment relations (OER) and market-oriented employment relations (MER)*

OER	MER
Employment Recruitment of workers to become members of an organization, with entry from the bottom and internal progression.	*Employment* Recruitment of workers to perform specific jobs. Entry at any point.
Payment Based on attributes and performance relative to the members of the organization. 'Market' references limited to starting rates and average rises.	*Payment* 'Market' rate as references for jobs or persons doing jobs. Job or skill relativities within the firm.
Industrial relations Contoured to the organization according to organizational wages and internal career progression.	*Industrial relations* Contoured along skill or occupational lines according to skills being bought and sold, and external labour markets.

greater flexibility in moving workers between jobs without reference to pay;

recognition and evaluation of the contribution of the individual towards organizational goals beyond narrowly defined production-related performance;

greater reflection of company performance in payment, since all are assumed to be 'in the same boat';

greater harmonization of employment conditions for all regular members of the company.

Regarding industrial relations, with OER relative to MER there would be:

extensive use of joint consultation and other communications channels to foster 'organization consciousness';

a focus of industrial relations and personnel management at the same level as business and corporate planning functions, the former being an integral part of the latter;

limits of principal worker organization coinciding with those of the employing organization;

a greater sense of 'common destiny' between managers and workers, with a blurred dividing line promoting this.

The MER pole represents the archetypal contractual relationship with minimal overlap of interests beyond immediate economic ones. There is a clear dividing line between those selling their skills and those buying them, and compliance, in Etzioni's terms, is gained through remuneration and sometimes coercion. Attempts at normative compliance will be greeted by scepticism as being irrelevant to the basic nature of the relationship. There is an 'Us–Them' relationship between those selling their skills and those buying them.

The OER pole, on the other hand, represents a relationship with a social as well as an economic dimension (a more 'diffuse' relationship in Parsons' terms). The relative unimportance of specific job-related pay and the norm of upward mobility blurs the employment contract line, and the preferred means of gaining compliance is through normative integration. There *is* a coercive element in the relationship, because if a worker is not cooperative, his[8] long-term pay and promotion prospects will suffer, but the coercive element is muted because (1) too much reliance on overt coercion is seen by managers as detrimental to the attainment of organizational goals they have established, and (2) the future of the individual becomes progressively more tied up with that of the organization, anyway. The 'Them' is externalized to those in competitor firms.

If 'Us' is to apply to most or all members of an organization, managers must be seen to be acting in the same interests as the workers (or at least convince workers that their interests coincide), and not as agents of parties on the other side of a market relationship. The power of outside shareholders to influence organizational goals and activities in their favour, therefore, is less in OER than MER, which implies a different *de facto* if not *de jure* relationship among shareholders, managers and employees. The MER firm belongs to its members – those who provide the capital and purchase the labour or labour power of the workers – while the OER firm 'belongs' to a significant extent to those who work in it, who are *its* members.

The ideal-typical constructions OER and MER also have something in common with formulations such as McGregor's (1960) Theory X and Theory Y, and other categorizations of managerial beliefs or strategies – Littler's has already been mentioned – but they involve often institutionalized relations which are not only created by these beliefs and strategies, but which also help to create them. There are some similarities with Williamson's (1975) Markets and Hierarchies modes of contracting, but while the employment relation *is* the hierarchy, and distinct from a contractual market relation for Williamson, here MER refers to a contractual form of employment relation. Since Williamson does acknowledge different degrees in interfirm market relations, presumably his framework would also allow for different degrees of hierarchy. In both conceptions the hierarchy or organization orientation is associated with a higher degree of cooperation with more scope for sequential decision making, while the problems of opportunism more frequently arise in the market mode or orientation.[9]

Fox (1974, 71–4) also has cogently described the link between relationships which are fundamentally contractual in nature (MER here) and spirals of mistrust.[10] The minimal overlap of interests and the dichotomous buying/selling relationship gives rise to these, as both sides seek to maximize their returns. The institutions of OER, however, encourage competition against an outside 'Them' to enlarge 'the pie' rather than disputing the division of it with an internal 'Them' (with little penalty in the latter for seeking out another pie to divide if necessary). The employment, pay and industrial relations of MER referred to here would correspond quite well with Fox's 'institutionalized mistrust' which he mentions but does not elaborate on, and OER to 'institutionalized trust'.

There are two respects, then, in which the influence of employment relations on the use of new technology should be discussed; the first in the context of practices logically deriving from the employment relations themselves, and the second in the context of practices brought about by the differing degrees of cooperation and conflict the employment relations engender.

Britain and Japan

While no company is likely to have either pure market-oriented or pure organization-oriented employment relations, there is reason to expect that more Japanese companies will be towards the OER end of the spectrum, and more British companies towards the MER end, even though non-manual workers may be employed under OER or semi-OER in the latter.

Dore's (1973) study of Hitachi and English Electric (two factories each), from which the terms 'organization orientation' and 'market orientation' derive, points in the same direction. Many of the features of the respective orientations are related to the period in which industrialization started, according to Dore; the 'late development effect'. Industrialization started out with small firms in the earliest industrializer – Britain – which, coupled with the market philosophy, shaped labour market and industrial relations practices. While there was a small firm sector started by indigenous entrepreneurs in an open type of labour market (or family-oriented or paternalistic) in the later developer Japan, there was also an early large firm sector which was important in the development of labour market and industrial relations practices.

Large firms represented significant capital investment, and governmental concern and influence was strong. They were big enough to develop internal career structures, as in government employment, their stability enabled them to guarantee employment to their employees, and more effort was made in developing communication channels. Coupled with this in Japan was the desire to avoid the worst conflict associated with contractual relations in the earlier developers. These differences in starting points were important because 'organizations tend to preserve features characteristic of society at large at the time of their foundation' (Dore, 1973, 138).

Dore's theory would predict a variety of employment relations in Japan, especially between large and small firms, and some movement towards organization orientation in Britain, although not necessarily even or unidirectional. Of course the development of employment relations in both countries cannot be divorced from the specific historical and sociocultural contexts independent of social actors. Certain aspects of craft organization in Britain, for example, precede industrialization and factories, while employment relations in Japanese industry were markedly more market-oriented at the beginning of this century than they are today. World War II and its tumultuous aftermath were very important in the shift to organization orientation. These developments and employment relations in both countries will be discussed further in chapter 2.

1.3 Employment relations, computerized machine tools (CNC) and skills: four hypotheses

Below are four hypotheses concerning the relation between employment relations on the one hand, and computerized machine tools (CNC) and skills on the other. They are based on a sizeable literature, and each hypothesis is followed by a brief summary of supporting arguments.

Hypothesis 1 concerns training, which is related to operator skills, while Hypothesis 2 concerns task ranges, which are related to skills a job requires. Hypothesis 3 concerns the relative influence of employment relations and factors such as factory and batch size on CNC use; are employment relations a major or a minor influence? Finally, although skills are discussed in the first two hypotheses, these are combined in Hypothesis 4 to facilitate a general and summary discussion.

Hypothesis 1: More training is given to CNC operators where employment relations are organization-oriented than where they are market-oriented.

Training can refer to a number of things. With respect to CNC operators, it can be broadly divided into two different categories; training over time in various jobs, which forms the basis of machining skills but is not specifically directed at operating a certain CNC machine, and training specifically for CNC operating.

Where internal careers predominate (OER), one might expect these two types to merge, and for CNC operating to be slotted into a hierarchy of jobs of increasing difficulty and responsibility, as in figure 1.1. Many of the skills necessary for CNC operating, in this case, will be imparted on the daily job.

On the other hand, if an individual is employed for a specific job which is not necessarily seen as preparation for the next job in a progression (MER), more training may have to be given at the changing of jobs, possibly outside the company. If the onus is on the individual to compete to get a new job, more of this specific training will be done at his or her initiative.

Training might also be organized into one extended period, such as an apprenticeship, in which apprentices would forfeit higher immediate wages for higher eventual returns. In either case, workers may actually receive more training, since it is less streamlined for specific jobs required.

On the other hand, broader job experience may be given in OER to promote flexibility and greater understanding of the relationship of different jobs in the work process, as Koike argues is the case in Japan (1977, 1981b), and human capital theory would predict that where managers are afraid of losing workers to external markets (MER) they will be reluctant to provide training – at least at their expense.[11]

In a contractual relationship where payment is made in principle for the job(s) done or the ability to do a certain job or jobs, someone with higher training would presumably try to extract the maximum price for his/her skills, while from the company side labour costs would be minimized by employing the minimum combination of skills necessary to get jobs done. *In theory*, then, equilibrium would coincide with minimum possible skill deployment, which implies minimum possible training.

OER payment, however, does not depend on the particular job being done. Labour costs might be minimized by using younger employees, but if it is not easy to make employees redundant given the norm of long-term employment, older, experienced workers would end up performing simpler tasks elsewhere (which would be against the norm of increasing responsibility). In this situation the career progression approach would be the most logical, and there would not be the same pressure to employ the minimum amount of skills necessary.

Furthermore, there is always bound to be a certain amount of bounded rationality involved in judging the amount of training necessary for new tasks. Managers would tend to choose the higher end of the scale who were (1) particularly concerned with quality, which is not necessarily related to employment relations, or (2) particularly concerned with morale, which *is* related to employment relations, particularly, as was argued, to OER.

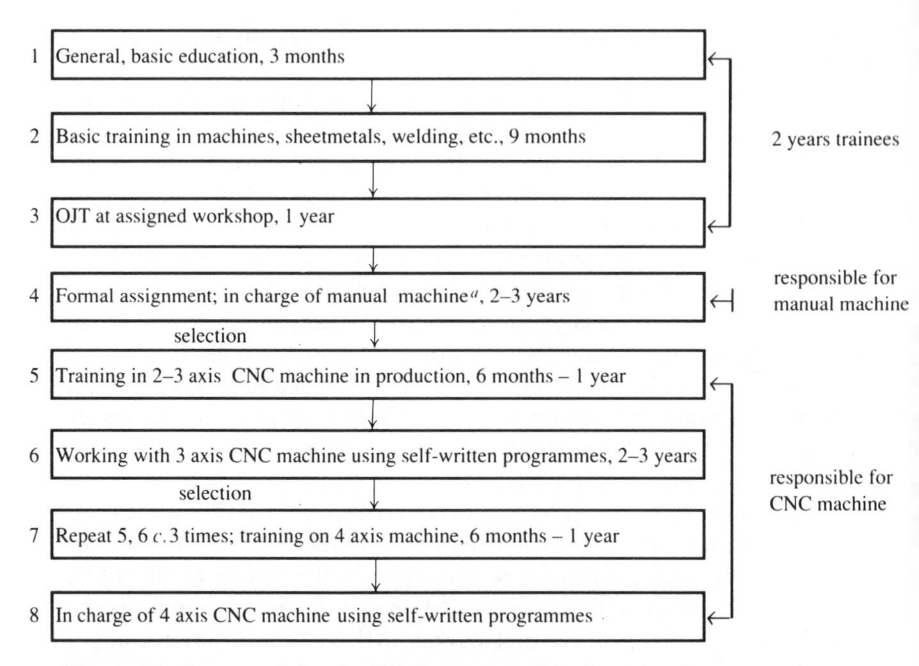

Figure 1.1 Career training for CNC operators in a Japanese factory
[a]Manual machine refers to a manually controlled or conventional machine tool.
Source: Tsusansho ed., 1984, 161

Finally, as Francis (1986) noted when discussing the deskilling debate, there are ways other than classical exploitation of labour and deskilling workers to achieve profitability – such as increasing market share, etc. In an OER situation, attempts to introduce new technology to supplant workers and skills (as opposed to augmenting these, or for handling 'dirty' work) would undermine the preferred method of gaining compliance – normative – hence would be avoided where possible. Of course introducing new technology in such a way would arouse fears if not outright opposition where remuneration is the main means of gaining compliance, and in keeping with this control mechanism, it might have to be sold to workers to ensure smooth introduction in the form of premiums. Using Fox's terminology, however, it seems likely that the avoidance of low-trust initiatives will be higher on the list of OER managers' priorities than those in MER. If control may become an end in itself on the one hand (see Noble's account of the introduction of NC at GE's Lynn factory; Noble, 1984), avoidance of low-trust initiatives and delegation of responsibility may be an important consideration on the other.[12]

Hypothesis 2: A wider range of tasks is performed by CNC operators where employment relations are organization-oriented than where they are market-oriented.

CNC tasks can broadly be divided into three clusters: operating which includes button pushing, machine minding and often loading and unloading the machine; setting of tools and/or the workpiece which, depending on the machine, workpiece and operation, may require considerable skill and may be done by a specialist setter; and programming, from proving out and editing tapes to full programming, which is at the heart of the deskilling debate. An operator may also perform related tasks – while a part is being machined, for example – such as inspection, deburring rough edges, preventative maintenance, and so on.

There are at least two potential obstacles in MER factories – besides workers' fears of labour intensification – to extending an operator's task range. First, the extended task range may encroach on to the job territory of others, who might be in a different union (see Jones, 1982, for an example of this with operators and programmers of the then AUEW and TASS). Not only programmers, but also such groups as slingers (doing loading and unloading) might be threatened if the task range of operators is extended. Secondly, operators may demand more money for more tasks, particularly where they are organized and have the power to press their demands.[13] This could easily upset existing relativities, hence there is a double potential for conflict.

This should not happen in an OER situation because the individual is not employed to do specific tasks, and pay is not specifically related to the tasks or number of tasks done. Also, worker organization is contoured to the employing organization rather than to occupation, hence territorial encroachment would not create an industrial relations problem, although it might trigger individual or departmental defence of territory. Thus there are fewer institutional constraints on expanding operators' task ranges, and there are incentives for workers to cooperate in terms of long-term wage and promotion prospects. This argument finds support from Dodgson's (1985) study of small firms in Britain using CNC, in which he found that where pay systems rated the person, more tasks were performed by CNC operators than where they rated the job. These in turn were linked to the presence or absence of unions – in other words, the degree or strength of MER-type industrial relations.

Regarding the link between MER and low trust and between OER and high trust, Fox (1974) argues that there is a dynamic relationship between low trust and low discretion in jobs. A lack of trust leads those higher up the organization to limit the discretion of those further down. Those mistrusted resist this, which provides further evidence that they cannot be trusted, and so on. Where there is a dynamic of low trust, CNC operators can be expected to perform a narrower range of tasks than where this dynamic does not exist.

Braverman emphasized the prising apart of conceptual tasks from execution, or programming from other operating functions. Such Taylorite strategies result in narrower and narrower task ranges for shop-floor workers. Braverman has been criticised for seeing Taylorism as *the* mode of capitalist control (e.g. Friedman, 1977, Wood, 1982), and some researchers claim to have found evidence of 'reskilling' and extended task ranges rather than 'deskilling' and narrow task ranges. A wide-scale survey in Japan, for example, found an initial polarization of skills as production engineers were made responsible for new machines and carried out programming, followed by a depolarization as those tasks were passed on to the operators (Koyo shokugyo sogo kenkyujo, 1985b, 18).[14] An earlier study argued:

> The fact that Japanese workers perform a wide range of jobs and show surprisingly high adaptability to changes may be due to the fact that skilled workers' potentials have been developed and they are well motivated. The underlying factors include (i) the existence of many employees who have the sense of belonging to the community bound together by common fate, and identify their own interest with that of the company in Japanese society where life-time employment is the rule, (ii) the maintenance of competitiveness among employees at a high level in the community of equality and uniformity, (iii) the insuring of the improvement of a wide range of skills and the growth as a human being through work. (NIEVR, 1983, 26)

Once again, if 'deskilling' is a managerial priority, this is likely to be in MER factories, while 'reskilling' and extended task ranges will be more characteristic of OER.

Hypothesis 3: The influence of factory size and batch size on CNC use is mediated by employment relations.

A study of 12 factories in West Germany and Britain calls into question the arguments being developed here. The researchers found that:

> In both countries, operators retain setting and programming-related functions to a greater extent only when the batch size is smaller. Larger batches pervasively mean that operating is more differentiated from preparatory functions, both planning, programming and setting. Large plants are more likely to have

> separate planning and programming departments. Small plants have particular programming and setting employees, but their tasks and authority are more used in personal patterns. (Sorge *et al.*, 1983, 95–6)

In other words, CNC use could best be predicted by knowing the factory size and batch size rather than particular management strategies or even employment relations. Country differences were related to approaches to training and production organization rather than to union strategies or demarcation disputes.

The study raises several questions; (1) to what extent can these findings be generalized – representativeness; (2) do factory size and batch size largely determine how CNC is used, or must one refer to other variables such as employment relations – degree; (3) are the differences best explained by reference to production organization and training, or are these themselves shaped by employment relations – national differences. The last of these will be discussed later in the book; the first two are relevant here.

Regarding factory size, large factories will indeed have separate planning and programming departments or sections, but it makes little difference to the operator's task range whether the tasks are done by a specialist *on* the shop-floor or *off* it. In fact, the presence of someone on the shop-floor responsible for programming might inhibit operators from doing such tasks as prove-outs or editing. From the *operator's* point of view, then, the effect of factory size is by no means unequivocal. Secondly, the NIEVR survey referred to above claimed that operators in larger factories were highly skilled and performing a wide range of tasks – more so than those in smaller factories (NIEVR, 1983, 7). This was due to policies stressing employee development, and greater resources to achieve it.

The effect of size may therefore differ according to employment relations; in large OER factories, where there is more scope for internal careers, training and socialization, size might indeed correlate positively with task ranges whereas the correlation might be negative in MER factories. In other words, the size factor has different effects depending on the employment relations context, while employment relations are influenced by size.

Coming to the effect of batch size, which Sorge *et al.* found was associated with a polarization of tasks, again there are contradictory findings in Japan, where operators had no input into programming because of small and extremely complicated batches (Koyo shokugyo sogo kenkyujo, 1983, 7–8, 40–1). As Sorge *et al.* were told, a CNC machine is too expensive to be used as a programming console (1983, 150), and a machine is likely to be idle more where batch sizes are smaller. What may explain their findings, however, is that with large batches, setting tasks may already be separated from operating tasks, and in this case, it is unlikely that operators will be given programming to do. It may be the case, however, that with different employment relations more setting will be done by operators – for motivational reasons, for example – even where batch sizes are large, and programming tasks will be added to these.

Again, the batch size might have different effects depending on the employment relations and vice versa, and it will be difficult either to predict or interpret CNC use without reference to employment relations.

Hypothesis 4: Skill levels are preserved or enhanced where employment relations are organization-oriented, and contested where they are market-oriented.

The word 'skill' has not yet been defined, and if 'skill' can refer to a number of things, so can 'deskilling' and 'reskilling'. Skill can refer to the cognitive and manual capabilities possessed by an individual which enable him to carry out certain tasks even though he might not be doing those tasks all the time. These may be called *operator* or *substantive skills*. It can also refer to the cognitive and manual capabilities necessary for certain tasks to be performed, independent of the actual substantive skills of the individuals performing them. These may be called *skill requirements*. Skill can also refer to the level of cognitive and manual competence groups of workers claim to have attained, or alternatively that managers claim workers need, in order to carry out tasks classed as skilled, semi-skilled, etc., whether or not the workers actually have those skills or require them to carry out the indicated tasks. These can be called *skill labels*.[15]

While not ignoring labelling possibilities, the analysis concentrates on the first two meanings. Substantive skills are looked at in Hypothesis 1, which considers the training and backgrounds of CNC operators, and skill requirements in Hypothesis 2, which considers ranges of tasks performed. The hypotheses suggest that these are influenced by employment relations and correlate positively with the degree of OER. Substantive skills are often ignored in the deskilling debate, but as we shall see, are crucial to an understanding of CNC use.

Hypothesis 4 is in effect a rewording of the other hypotheses, and is intended to facilitate a general discussion and summary. This will include the relationship between substantive skills and skill requirements, with the expectation of a minimal relationship – the least substantive skills for given skill requirements – in factories nearer the MER pole. The main reasons for this have already been given, and derive from both the characteristics of employment relations and the degree of trust or mistrust that they engender. In MER these also lead to greater potential for conflict. Again, if there is any attempt at 'deskilling' workers or jobs, this is likely to be in factories nearer the MER pole, while 'reskilling' should be more characteristic of OER factories.

A number of arguments could be ventured against these hypotheses. One could argue, for example, that substantive deskilling is avoided by apprenticeships, and deskilling of jobs by shop-floor control – Hobsbawm (1984, 192) argues that this is extensive in British factories – and that this resistance is easier to maintain where industrial relations are contoured along craft rather than organizational lines. To this writers like Lee would respond 'neither at local nor at national level have craft unions been able to impinge successfully on employers' "right to manage" industrial training or to enhance the status or content of apprenticeship' (1982, 159). Dodgson's (1985) study of small British factories and Kelley and Brooks' (1988) large-scale survey of United States firms show greater task ranges where unions are not present.

Some Japanese writers claim 'deskilling' strategies to be widespread in that country; indeed, these arguments have a long history, preceding Braverman, although the motive is more of saving labour costs and less of control.[16] That does not represent the current mainstream of opinion, however, as suggested by earlier quotations. While the arguments for the hypotheses seem stronger than those against, care has been taken not to *impose* order and meaning on the work; it will be just as interesting if some or even all of the predictions made are not upheld.

1.4 Eighteen factories

Machine tools are used in a broad range of industries: by metal product makers of all descriptions, by parts makers, by specialist machine makers for in-company use, and so on. In some cases they are central to a factory's operations, in others peripheral. Some factories use machine tools to turn out large batches and others to turn out small ones. Some factories use computerized machine tools more extensively than others; some have integrated them into more advanced flexible manufacturing-type (FMS) systems, some use them on a 'stand-alone' basis (i.e. singly), while some do not use them at all.

In the factories selected here, machining had to be a central operation, and they had to have at least three CNC lathes, milling machines or machining centres (one exception was made for the smallest Japanese factory, which only had two), but not have advanced, integrated systems which would fundamentally change the job of the operator. As well as technology, the factories were matched as closely as possible by product, size and batch size (cf. Sorge *et al.*). Having differently-sized factories would

Table 1.2 *The 18 factories*

Factory	Employees	From	Company	Factory	Employees	From	Company
J1	7	1960	7	B4	35	1952	35
J2	19	1945	19	B8	81	1951	11,000
J4	42	1938	42	B11	110	1889	110
J9	86	1960	86	B12	115	1957	135
J45	435	1949	530	B39	390	1962	1,500
J50	504	1910	740	B71	709	1886	740
J66	657	1919	846	B145	1,450	1871	4,600
J140	1,400	1949	35,000	B80	800	1893	45,000
J180	1,800	1949	35,000	B120	1,200	1910	45,000

Factories making the same or similar products are J45 and B39; J50, J66, B71, B145, B11 and B8; and J140 and B80. J180 and B120 are elite factories in terms of engineering precision and skills in the same companies as J140 and B80 respectively, but their products differ. B4 used to make the same product as J4, but changed recently, as did B12, which was approached as a counterpart for J9.

also ensure a variety of employment relations (as the work of the Aston School, Mintzberg, etc. would suggest). The matching was reasonably successful, particularly for the larger factories.[17] Two very small Japanese factories, so-called *machi koba*, were included because the way they use CNC is reputedly different from the larger factories, and even from the other small factories of 50–100 employees. As we shall see, these very small factories are a significant entity in Japanese industry.[18]

The factories are listed in table 1.2. Some wished to remain anonymous. They are referred to throughout in terms of country and size. J refers to a Japanese factory and B to a British one, with the number following the letter showing the total number of employees divided by 10 and rounded off. Thus, a Japanese factory with 657 employees is shown as J66. The founding date of the company with direct lineage is given in the third column (not necessarily the present owners, who might have acquired the factory or company recently), and the total number of employees in the company in the fourth.

Table 1.4 shows changes in employment over the period 1977–87. Only two of the British factories had grown in size while most, particularly the larger ones, had experienced considerable reductions in their workforces. Only three of the Japanese factories had seen reductions in their workforces, none of them as drastic as in the large British factories. The figures are indicative of the well being of the respective mechanical engineering industries.

Table 1.3 *The 18 factories: batch sizes*

Factory	Smallest	Largest	Average	Factory	Smallest	Largest	Average
J1	1	20	4–5	B4	1	50	6
J2*	40	110,000	2,000	B8	1	100	6–10
J4	1	20	4–5	B11	1	300	10–15
J9	1	12	3–4	B12*	1	10,000+	500+
J45	1	20	2–3	B39	1	200	6–7
J50	1	4–5	1–3	B71	1	100+	15–30
J66	1	200	10–30	B145	1	200	6–7
J140*	90	5,000+	250	B80*	25	8,000	350
J180	1	100	3–5	B120	1	100	4–10

*Indicates large batch factories

Table 1.4 *Changes in employment, 1977–87*

Factory	1977	1982	1985	1987	Factory	1977	1982	1985	1987
J1	15	12	9	7	B4	39	37	35	35
J2*	12	15	18	19	B8	105	95	95	81
J4	12	28	40	40	B11	117	115	115	110
J9	50	65	72	89	B12*	95	110	110	115
J45	400	410	450	450	B39	380	380	400	390
J50	650	600	550	500	B71	1,100	960	930	710
J66	710	690	650	660	B145	3,170	1,750	1,590	1,450
J140*	1,000	1,300	1,400	1,400	B80*	2,400	2,100	1,700	800
J180	1,000	1,350	1,650	1,800	B120	1,900	1,770	1,080	1,200

*Indicates large batch factories

Certain types of factories were avoided as being likely to present a biased picture, for example machine tool makers, whose factories often serve as showrooms. There is a tendency particularly with studies of Japanese industry to generalize observations of technology leaders, resulting in a distorted picture of those which are not. Although some of the factories were leaders in their field, the overall selection should give a more balanced view of the mechanical engineering industry in both countries, although complete representativeness obviously cannot be claimed. More details of the individual factories, and the way in which the information on them was gathered, are given in appendix 2.

2
The wider context

One of the main reasons for looking at Japanese and British factories lay in the reputations both countries have for contrasting types of employment relations. If reputations are anything to go by, more Japanese factories, with their 'three pillars' will be nearer the OER pole, and more British factories, with a tradition of 'individualistic market contractualism' (Fox, 1985a) will be nearer the MER pole. Reputations can be misleading, however, as some writers suggest. In the first section of this chapter I will look briefly at grounds for these propositions, and also at whether or not employment relations have helped the introduction of new technology in Japan more than in Britain as some have claimed, citing macro-level studies.

How are the differences (and similarities) best explained? The second section looks at various approaches to employment relations and their limitations, including culture, history, ownership and power relations. This will provide background information which will help put the later chapters in perspective. Finally, there are recent signs of change in employment relations in both countries, ostensibly in an OER direction in Britain and a MER direction in Japan. The last section looks at these.

2.1 'Lifetime employment' and 'market contractualism': a macro view

Employment

The 'three pillars' formulation of Japanese employment relations has come back under attack in recent years. 'Lifetime employment' in Japan is a myth, some have claimed (e.g. Koike, 1983, 1981b, see also Levine, 1983). Koike argues that lengths of employment in Japan are not very different from those in the United States. In fact, older workers in Japan have had *less* continuity in employment than their US counterparts.

In a recent study, the Economic Planning Agency of Japan tried to ascertain the extent of 'lifetime employment' in that country. It allowed in

Table 2.1 *'Fixed employment' in Japan and Britain (all sectors)*

	Male		Female	
Age	Japan	Britain	Japan	Britain
30–4	78.5	61.1	63.7	34.8
35–9	69.6	46.3	35.6	16.2
40–4	62.0	54.1	18.0	23.4
45–9	50.9		10.6	
50–4	36.5	35.2	7.7	10.7
55	17.5		5.0	

Sources: Japan: *Chingin kozo kihon tokei chosa*, 1986; Britain: Department of Employment, 1986 (data base).

its calculations for higher levels of turnover of those under 30 years old. Of junior high school leavers, for example, it noted that 63.5% leave their first place of employment in the first three years, while the corresponding figure for high school leavers is 35.8% (Keizai kikakucho, 1986, 110). 'Fixed employment' was defined as those workers meeting the criteria below. Fortunately, it has been possible to create similar categories from figures obtained from the Department of Employment in Britain, although certain categories are not exactly the same, and inflate the British figures:[1]

Age	30–4	35–9	40–4	45–9	50–4	55+
Japan: continuous employment (years)	5+	10+	15+	20+	25+	30+
Britain: continuous employment (years)	5+	10+	10+		20+	

Table 2.1 shows that a greater percentage of Japanese workers came under 'fixed employment' than their British counterparts in all categories except for older women, and it would seem that were the figures exactly comparable here, Japanese women would also have a higher rate of 'fixed employment'. Unfortunately, establishment size in Britain was only broken down into those with less than 25 employees and those with 25 or more, and the figures have been omitted here, but the largest British size category approached only the smallest Japanese (company) size category of 10–99 employees.

Table 2.2 shows that lengths of employment for British managers are not much greater than those of craft workers. In fact, 45% of professional managers and administrators had been with their establishment for less

Table 2.2 *Employment length by occupational category (Britain, all sectors, %)*

Full years in company		0	1–4	5–9	10–19	20+
Occupational group						
Managerial and professional	M	11.0	24.3	20.7	26.1	17.7
	F	16.5	31.5	22.0	22.3	5.6
Clerical and related	M	18.0	26.7	20.3	22.2	12.6
	F	20.3	34.4	23.5	17.6	4.1
Craft and similar	M	13.2	26.6	21.9	23.3	14.9
	F	20.7	36.0	21.5	16.7	5.5
General labourers	M	24.8	25.3	24.7	16.7	8.2
	F	35.8	27.0	13.4	13.8	10.0
Professional, managerial and administrative	M	14.7	25.3	24.7	16.7	8.2
	F	16.6	36.8	22.6	19.3	4.7

Source: As for table 2.1, all sectors. M = male, F = female. Figures are rounded, and those for 'other non-manual' and 'other manual' omitted.

than five years. Unfortunately, similar figures by occupation are not given in the Japanese survey, but differences between manual workers and non-manual workers can be calculated for both countries, and are shown in Table 2.3. In both countries, manual workers are more mobile than non-manual workers, but the differences in Japan are more striking.

Overall, the figures suggest that 'fixed' employment does characterize a greater proportion of Japanese workers than British workers, allowing for mobility of those under the age of 30. Of course, length of employment by itself does not demonstrate organization orientation or market orientation. A lack of alternatives, rather than particular types of employment relations or personnel policies which influence employment relations, might give rise to longer employment. The figures do not tell us anything about the rigorousness of the recruitment or selection process, about orientation training, or subsequent movement within the company. It will be more appropriate to discuss these when we look at the 18 factories, however.

Payment systems

Koike also notes (1988, 1983) that there is nothing unique about *nenko* wages as seen through age-earnings profiles. Those of blue collar workers in Japan are very similar to those of white collar workers in western countries and are closely related, he argues, to skill formation. As has been noted, some employees of a firm may be treated by OER principles and others by

Table 2.3 *'Fixed employment': manual and non-manual workers*

	Manual				Non-manual			
	Male		Female		Male		Female	
	Japan	Britain	Japan	Britain	Japan	Britain	Japan	Britain
Age								
30–4	79.4	59.7	54.0	23.8	84.7	62.4	70.5	40.3
35–9	70.5	37.3	26.1	9.1	82.0	48.5	48.6	19.7
40–4	59.5	48.7	11.2	17.4	78.7	58.8	29.3	27.1
45–9	43.8		5.7		72.9		20.3	
50–4	29.5	31.6	3.6	8.3	59.0	39.3	17.2	12.6
55+	18.7		2.8		33.0		13.6	

Sources: As for table 2.1. Japanese figures are for manufacturing, British figures are for all sectors.

MER principles, and our interest in this study is particularly with manual workers, where differences will be most apparent. We can expect less difference between the age-earnings profiles of non-manual and manual workers in OER relative to MER because in the former manual worker profiles will represent not only market rates for manual, skill and dexteral abilities, but also such factors as length of employment and organization-specific criteria, and also because harmonization of employment conditions is in all probability a prerequisite for fostering organization-wide organization consciousness.

Figure 2.1 shows the age-earnings profiles of manual and non-manual workers in Japan and Britain, with the highest category of earnings of non-manual workers in both countries taken as 100. The gap is indeed smaller for Japanese manual workers: the curve is not as flat, and the peak is further to the right. This is even more pronounced in larger Japanese firms; in manufacturing companies of over 1,000 employees, the earnings of manual workers peak over the 45–9, 50–4 periods at 77.9% of non-manual workers, whose profile peaks in the 50–4 bracket.

Age-earnings profiles do seem to indicate greater organization orientation in Japan than in Britain. As with 'fixed' employment, however, they do not tell us very much about the nature of the pay systems and how these are linked to OER and MER. They do not tell us the basis of calculating wages for manual and non-manual workers, if wages rate the man or the job, if workers are evaluated or not, if wages vary according to company performance or not, and so on. These are discussed in chapter 3.

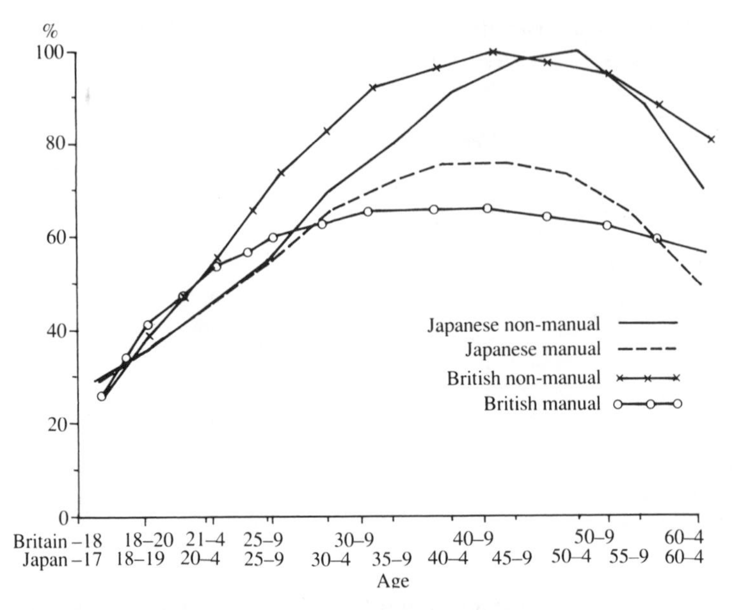

Figure 2.1 Age–earnings profiles in Britain and Japan
Note: Figures are for average gross weekly earnings for Britain and average gross monthly earnings for Japan. Highest non-manual figure for each country = 100.
Sources: Britain: NES, 1986, E65, 66; Japan: *Rodo tokei yoran*, 1986

Industrial relations

Tabular comparisons of industrial relations as they relate to our discussion are not particularly helpful. 'Enterprise unionism' may be the norm in Japan, as shown by table 2.4, but as Koike (1988) has argued, super-enterprise levels of worker organization are also important, as has been the *shunto* (spring wage offensive) for setting wage rises in individual firms. Super-enterprise unions and federations are especially important for workers in smaller enterprises, which have neither the resources nor the expertise of the larger enterprise unions at their disposal and hence rely more on paid outside officials (Toritsu rodo kenkyujo, 1983).

At the same time, the recognized focus of industrial relations in Britain has been changing. Sisson and Brown wrote 15 years after Donnovan that 'so far as the great majority of workers in the private sector are concerned . . . multi-employer bargaining can no longer be described as the formal system' (1983, 140). Multi-employer agreements have increasingly taken the form of a 'safety net' while the most important level of bargaining, at least in mechanical engineering, has become the individual establishment.

Table 2.4 *Membership by type of union (Japan)*

Year	Enterprise union	Craft/Occupation	Industrial union	Other
1964	91.4	0.7	4.9	3.0
1975	91.1	1.3	5.4	2.1

Source: Seisansei honbu, 1987, 160.

There are three main areas of difference regarding the organization of industrial relations (independent of the content) in large British and large Japanese manufacturing establishments or firms. First, multi-unionism is still predominant in Britain, and while representatives of manual unions may coordinate their activities (joint shop steward committees for manual unions were reported in 33% of private manufacturing establishments in the 1984 BWIR survey), there is still very little coordination between manual and non-manual unions. By contrast, although there are second unions in some Japanese companies, most have only one union, which organize both manual and non-manual workers. Where there are second unions, they are not normally differentiated according to occupation or even to the class of worker they organize (e.g. manual as opposed to non-manual), although one union might be stronger in certain areas than others. They are normally competitors, differentiated by degree of militancy and outside affiliation.

Secondly, the focus of industrial relations in Japan is much more concentrated at the *company* level. This is promoted by a tradition of internal growth and concentration in specific industries (see Clark, 1979), whereas where growth is through merger or acquisition – as has often though not always been the case in Britain – the incorporation of different traditions and industries mitigates against such concentration.[2]

Thirdly, whereas in Britain union membership is of a national union, in Japan it is normally of an enterprise union. Funding patterns are thus very different, as Dore (1973, 134–6) has noted. Fees in Britain go to the formal union, separate from the steward organization, while in Japan they are collected at the company level and a certain amount given to industrial or national federations if the union is affiliated. Even Zenkokukinzoku (National Trade Union of Metal and Engineering Workers), with an extensive national and district organization, was trying in the late 1970s to alter the union:district:(enterprise)branch funding ratio of 8:20:80 to 15:25:60, but only tortuous progress was made (Toritsu rodo kenkyujo, 1983, 30–1).

Not only are funding patterns different, but the amount (these vary in both countries according to the union) is two or three times greater per member in Japan than in Britain. Consequently, union representatives within the Japanese company have much greater financial resources at their disposal with which to invest in facilities, surveys, written communications materials, etc., and the inclusion of non-manual workers tends to promote these kinds of activities.

The thrust of these arguments is that even through union activities in Japan, consciousness of the company or organization is promoted, whereas the union organization in Britain can reinforce occupational, establishment or class identification. In other words, industrial relations in Japan is closer to the OER pole. We have, of course, barely scratched the surface of industrial relations. We have not considered, for example, informal channels of communications (34% of the BWIR survey establishments had a joint consultation committee in 1984; 72% in Japan in 1986) or the contents. Nor have we considered in any detail smaller companies in both countries, which are likely to be closer in industrial relations than the bigger companies, depending on the normative (and economic) influence of the latter in their respective countries. Again, the picture will become clearer as we discuss individual factories, which is the main level of focus in this study.

New technology

Many writers have argued that employment relations in Japan facilitate the uptake of new technology. The 'corollary' is that those in Britain have impeded it. Here I will take a broad look at industrial relations and the *introduction* of microelectronics technology; later we will look more carefully at its *use*.

It is widely accepted in Japan that cooperative employment relations resulting from the 'three pillars' have facilitated the smooth introduction of new technology (e.g. Shirai, 1986, 176; Itami, 1987), and successful technological innovation in turn is seen as a major factor – if not *the* major factor – in post-war economic growth (e.g. Yakabe, 1984). In contrast, 'western style' employment relations are held to impede smooth innovation, including organizational innovation (Urabe, 1984), and in the 'extreme case' of Britain, 'in spite of technological innovation, if solid, fixed rules govern the deployment of people, investment may be inhibited' (Koike, 1977, 5). On the other hand, Inagami (1983) notes the similarities in the basic positions of unions and federations in Japan and those in western European countries.

In Britain, the basic position of the TUC regarding microelectronics innovations is laid out in the report *Employment and Technology*:

> Technological change and the microelectronics revolution are a challenge, but also an opportunity. There is the challenge that the rapid introduction of new processes and work organization will lead to the loss of many more jobs and growing social dislocation. Equally, however, there is the growing realization that new technologies also offer great opportunities – not just for increasing the competitiveness of British industry but for increasing the quality of working life and for providing new benefits to working people. (1979a, 1)

The report urges that 'the priority for industrial policy is . . . to increase the rate at which technological advances are adopted by industry' (1979a, 16). In the Congress of 1979, fears about unemployment were voiced, but what was emphasized was 'the unique and unparalleled opportunity . . . for Britain to improve its economic performance and also its competitiveness in world markets while improving living standards, affording more leisure to employees, eliminating dreary work and improving communications between people and nations'.

New technology was also linked to the debate about industrial democracy; innovation was to be carried out with the full involvement of trade unions, formal agreements were to be made, and there should be right of access to information systems. There was a ten-point checklist for negotiators: no unilateral introduction, inter-union organization, the development of expertise, prior access to information, redeployment, retraining, reductions in working hours, no disruption of pay levels and structures, a move to equal conditions and single status, continued control over work, observance of health and safety guidelines, and review procedures to be set up.

The position of Denkiroren (Japanese Federation of Electrical Machine Workers Unions), one of the Japanese federations most favourably disposed towards technological innovation, is very similar:

> Our fundamental consideration regarding the microelectronics revolution is to bring about the 'humanization of work'. Our strategic goal is to gain a fair distribution of profits (gains) as the result of productivity increases through microelectronicization. To accomplish these we have eight concrete policies; a shortening of working hours, improvements in working conditions and the working environment, policies to bring about shop-floor development, training for technology and skill changes, policies for health and safety, for older and female workers, and strengthening the review procedures for technological systems; and three great principles:

> i) the thoroughgoing observance of prior consultation,
> ii) non-recognition of redundancies accompanying microelectronicization, and iii) thorough checks regarding health and safety. The pivot for carrying out these strategies is in the thorough observance of labour-management consultation. (Quoted in Inagami, 1983, 60)

The stances of Sohyo (General Council of Trade Unions) and the former Domei (Confederation of Labour, which merged with other private sector confederations in November 1987 into Rengo, or Japanese Private Sector Trade Union Confederation), representing basic continuity from earlier phases of innovation, have stressed a shortening of working hours, no redundancies, prior consultation regarding transfers, health and safety provisions, and full training, although Sohyo's initial policy was to achieve these through 'workshop struggles' whereas Zenro (forerunner to Domei) advocated prior consultation through production councils. In practice, however, the stances have not been radically different; in 1983, for example, the four national centres presented unified demands to the prime minister regarding the 'microelectronics revolution'.[3]

If official union responses have been no less positive in Britain, what about responses at the workshop level? Daniel (1987, 34) notes from the 1984 BWIR survey that there was certainly 'no sign that levels or forms of trade union organization at manufacturing workplaces inhibited the use of advanced technology in either processes or products'. In fact:

> Technical change was generally popular among the workers affected, but both shop stewards and full time trade union officers tended to support change even more strongly than the workers they represented. . . So great has been the support of workers and trade union representatives for technical change that managements have not had to use consultation, participation or negotiation to win their consent to change. (Daniel, 1987, 34)

Managers only consulted when required to do so, which was not so different from the situation in Japan; in 50% of the companies surveyed by the Ministry of Labour in Japan there was no consultation reported (Rodosho, 1985b, 197). Workers in the BWIR survey accepted advanced technological change even in the face of workforce reductions (more so than organizational change associated with employment stability: Millward and Stevens, 1986, 68–71), and were more positive about the new technology after it was introduced than before. Thus, there was no sign of greater worker or union opposition than, for example, in the JIL survey (1984) in Japan, although acceptance does not preclude serious bargaining.

As in Japan, one factor in this acceptance is that new technology has not been perceived by workers as leading to unemployment. New technology was well down on the list of reasons for reductions in the workforce in the BWIR survey, and natural attrition, redeployment within the establishment and early retirement were the preferred means of workforce reduction. (Further, capital investment is not poured into a company that is about to close.) Workers and unions seem to have been mindful of the threat of greater unemployment if competitiveness does not improve, and anxious to avoid being labelled Luddites.

One must be cautious, then, not only about associating British workers and unions with dogged opposition to new technology, but also about assuming that reactions to new technology are a simple reflection of employment relations. These reactions may differ from reactions to other kinds of change, and are shaped by a number of factors mentioned above. Whether or not this also applies to the *use* of new technology, as outlined in the hypotheses, remains to be seen.

2.2 Shaping factors

One explanation for the differences in employment relations between Britain and Japan was touched upon in chapter 1 – the period in which initial industrialization took place. A different approach, especially in Japan, has been to highlight cultural influences, which has triggered a spate of recent rebuttals. Employment relations – and interfirm relations and ownership, which influence them – evolve, and differ by sector. Evolution itself does not refute cultural influence; one could argue that greater organization orientation in Japan nowadays, compared to the period up to the 1920s, reflects a gradual process of indigenization of institutions and concepts originally derived from the earlier developers. It can be shown, however, that particular historical developments (including social, economic, political and technical changes) have shaped employment relations and ultimately these must be taken into account in a comparative discussion.

These issues are taken up in this section, in which various approaches to employment relations which might help explain the differences are discussed. We will start with culture and history, then look at sectoral differences within the two countries, followed by differences in ownership and finally back to culture, compliance and power relations.

Culture and history

Japan and 'Groupism'

> If Americans assume that all people are fundamentally the same and note differences as they appear, the Japanese assume that people are fundamentally different and note the similarities only as they appear. (Taylor, 1983, 40)

A number of Japanese scholars have written on the peculiarities of Japanese culture, institutions, interpersonal relations, and even of Japanese brains. The so-called *nihonjinron* (debate on the Japanese character) may be attributed partly to the lack of fit of 'universal' western theories and partly to self-defence against the invasion of western influences (and possibly also to the socioeconomic position of academics in the broader society; Mouer and Sugimoto, 1986). The intense criticism that *nihonjinron* have come in for is perhaps partly because of their potential xenophobic consequences, and partly because of the opposite assumptions of the critics that Taylor notes.

This discussion starts with one such variant, which not only attempts to explain the origin of Japanese-style employment relations, but also those of the 'west', including Britain. According to Odaka (1981), the formative influence on Japanese-style employment relations is *shudanshugi* (groupism) as against *kojinshugi* (individualism) in the west. The origins of *shudanshugi*, it is suggested, date back to cooperative rice planting communities of 1,500 years ago, whereas western individualism derives from hunting tribes (Germanic) and Christianity (which form is not clear).

Little evidence is advanced for these contentions, but they do raise the issue of continuity from the pre-industrial past. Arguments have been waged as to which of the four feudal classes – military, peasant, artisan or merchant – supplied the relationship type that became characteristic of modern Japanese employment relations. Sometimes the arguments are explicit, sometimes implicit, but some degree of continuity from feudal times is usually implied.[4]

These arguments often fail to specify the degree of generality of the relationship type or to demonstrate its relationship with specific institutions. Groupism and individualism, for example, are very general orientation concepts; how do they explain specific employment systems, payment schemes, industrial relations features, degree of participation, and so on? Do relationships in one type of institution tend to produce certain character types which are then influential in the creation of other institutions?

Japanese employment relations have evolved. Labour turnover was extremely high in the early 1900s – over 100% in many cases – and mobility was a central part of a 'proper' career (Gordon, 1985, 35).[5] As in Britain in the nineteenth century, manual workers were employed through the intermediary of a kind of foreman or master (*oyakata* in this case), then core workers came to be employed directly by the firm, but it was not really until the economically depressed 1920s that some form of 'lifetime employment' for core workers became established.

The motives were, from the managers' side, in part to incorporate or control the *oyakata*, but also to train workers for a finer division of labour determined by the company, to retain skilled workers, and to subvert the burgeoning labour movement. From the workers' side there was a search for security as workers with more education began to think in terms of 'careers' and 'prospects', while financial hardship also provided a strong incentive.

Early attempts were made by workers to organize by trade or industry or groups of companies, but with the combination of employer suppression and cooption, these attempts foundered in large enterprises, so that by 1925, there were more workers in enterprise-type 'unions' than other types (Sumiya, 1986).[6] Status differences remained sharp, however, and labour turnover was considerable, even during World War II, when it was officially frozen. White collar workers received monthly wages with large bonuses, while manual workers received daily wages with a very small bonus at the end of the year.

Government attempts to ensure full mobilization for the war effort also influenced the evolution of employment relations. Bureaucratic regulation attempted to reduce labour turnover, foster wages which took into account life cycle needs of workers, enforce systematization and reporting of wage systems, and promote a 'shared community of fate' ethos amongst managers and workers, which the *Sanpo* (literally 'industrial service to the nation') association and councils also called for. Gordon argues that broadly speaking, 'Japanese-style' employment is the result of pre-war management initiatives, wartime bureaucratic decree, early post-war worker pressure and management revision of post-war changes. An important factor throughout was the appeals or demands of manual workers not to be treated as second-class citizens.

After Japan's defeat in World War II, and with the subsequent occupational reforms, a blue-collar-led, invigorated worker movement led to a great reduction in status differences, with enterprise-based unionism (starting from large factories) incorporating both manual and non-manual workers serving as the vehicle for this. New wage systems such as the *densan* (electric power industry) system were established. These spread and were modified as managers regained the upper hand in shaping employment relations. Reduction of status differences, however, continued. New paths for promotion from the shop-floor were opened up. Human relations theories from the United States encouraged improvements in communications, and there was, of course, the importation of quality control concepts, all of which were modified in practice.

In brief, modern Japanese employment relations cannot be understood in separation from the historical context in which they evolved. As was suggested in chapter 1, the period of initial industrialization was important, and foreign influences have been critical, although these have been adapted. Culture has played a part in this adaption, but while in a broad sense 'groupism' and associated values might be significant, they cannot in themselves explain the specific institutions of employment relations which shape employer–employee interaction.

Britain and 'market contractualism'

The search for the roots of British employment relations, particularly contractual relations, has also taken researchers back hundreds of years. Although not specifically concerned with employment relations, Macfarlane traces English individualism back to feudal or even pre-feudal times:

> Those who have written on [the origins of modern individualism] have always accepted the Marx–Weber chronology. For example, David Rieseman assumes that modern individualism emerged from an older, collectivist, 'tradition-oriented' society in the fifteenth and sixteenth centuries. . . Yet, if the present thesis is correct, individualism in economic and social life is much older than this in England. In fact, within the recorded period covered by our documents, it is not possible to find a time when an Englishman did not stand alone. (Macfarlane, 1978, 196)

The author concludes his book by saying that his suspicions as to the time and place of the origins of English individualism are similar to those of Montesqueu (cf. Odaka also): back in the forests of Germany.

Martin (1977, 93–9) traces the shift from personal-diffuse to personal-specific to impersonal-specific relationships back to the twelfth century, which saw the revival of a monetary economy, population growth, the revival of cities, and the formalization of contracts. As contracts became formalized they became negotiable, the option of which became available to increasingly large segments of the population. The work of Fox also suggests continuity not only contractual but also of master–servant relationships from the pre-industrial past.

Furthermore, whereas the Webbs held that trade unions were a development of the late eighteenth to early nineteenth centuries, writers such as Leeson (1979) argue that their evolution can be traced back to thirteenth-century guilds. The history of the evolution of trade unions is one of the gradual alienation of journeymen and poorer masters from richer and more powerful masters who became more and more closely identified with merchants and merchant interests. Unions are not just a radical response to the industrial revolution and the factory system.

Of course employment relations in Britain have evolved, not only before but also after the industrial revolution (with an inordinate emphasis on control strategies according to Fox, 1985a, 52). The 1880s, for example, saw a movement towards eleminating the butty system similar to what later happened in Japan, the widespread introduction of piecework payment systems, the decline of the tramping system and the rapid growth of general union membership.

There were also possibilities for developments in other directions. There was some debate on 'livelihood' or 'life cycle' wages in the eighteenth and nineteenth centuries (e.g. Dobson, 1980), but the idea never gained momentum. There were welfare paternalists in the nineteenth century, but not enough to change the direction of development of employment relations (Dore, 1973). There were possibilities for change after World War I (as for Japan after World War II), but they were ultimately aborted by the triumph of the restorationists over the reconstructionists (Littler, 1982). There have been numerous developments since then – some related to scientific management in the interwar period, others to 'human relations' concepts, the growth of the shop steward movement and formalization of workplace industrial relations, the formalization of payment systems and decline of piecework in many industries, and so on – but these have done little to alter 'instinctive suspicion' and 'traditional ideologies' stemming from individualistic market contractualism according to Fox (1985a, b).

There *are* grounds for agreeing with Fox about the 'strength of the social forces which have shaped Britain's industrial relations over the centuries' (1985a, 429), and for arguing that there has been greater continuity in the basic nature of employment relations in the first industrializer, where industrialization and factories were largely the result of indigenous developments rather than externally imposed challenges. There are *also* grounds for agreeing with Brown that:

> A crisis of international competitiveness in the private sector, and of finance in the public, has forced major reappraisals of the way labour is contracted and motivated. The cumulative effect of many innovations is that collective bargaining is taking new forms and that trade unions, especially in the

> private sector, are being forced to adopt new structures.
> (Brown, 1986, 161)

We will look at these later in this chapter and in following chapters. Such concepts as 'individualism' and 'groupism' are informative at a relatively high level of abstraction, but the concrete features of modern British employment relations must be understood in their historical context, too, even if within a 'market contractualist' tradition.

Sectoral differences

Japan

The biggest weakness in arguments which attribute employment relations largely to cultural patterns lies in the significant differences by sector. Here we shall see that different influences operate on different sectors of the economy, resulting in different patterns of employment relations.

The public sector in Japan is not noted for its cooperative industrial relations. Unlike present disputes in the private sector, which take place within an accepted framework of industrial relations, those in the public sector have often concerned the framework itself. The militancy of public sector unions is attributed by those in the private sector to the lack of a profit motive – employees do not face the consequences of their own militancy – but there are also legal issues, restrictions or prohibitions on the right to bargain collectively, and especially the denial of the right to strike (which led to an ILO Commission in 1965). Furthermore, governments in the past often ignored the wage recommendations of the responsible third party commission (see Koshiro, 1979, 1983, Yamaguchi, 1983).

The ever-changing union Doro (National Railway Motive Power Union) in Inagami's (1981) account cannot be described as cooperative or consensus-seeking. Originally a conservative, elitist union, it gradually became more militant from around 1960, and was the most radical railway union after the *marusei* productivity drive of 1971 (only to become cooperative again in the recent privatization of the National Railways). There were two types of 'Them' for Doro members, the first being the hopeless, spineless, petty superiors, the second being members of competing unions, particularly the 'yes men' of Tetsuro and Zendoro. Group solidarity was promoted, as was an anti-promotion culture (to 'Them'), or promotion by strict seniority only, and there was mistrust and a lack of communication between 'Us' and 'Them'.

This horizontal solidarity, however, may have been partly a result of the employment system itself. Station masters of larger stations were graduates of prestigious universities, grade 1 public servants, high fliers who were there to get experience before being moved on, while assistant masters (grade 2) were graduates of less prestigious universities. Grade 3 workers – locally employed school leavers – resented younger university upstart 'outsiders' who often did not know how to construct a *daiya*, or schedule. The same might apply to employment in the post office network.

Apart from the public/private sector differences, there are significant differences within the private sector; between industries and between large and small firms. Roughly 70% of all employees in Japan are employed in the so-called 'small-and-medium sized enterprises' (hereafter called 'smaller companies'). In manufacturing, the figure is slightly more than 60% (table 2.5).[7]

The wages in smaller companies are lower than those in larger firms and welfare benefits correspondingly small (table 2.6). Many of the smaller companies are family owned and the name denoting this – *dozoku kaisha* – has a less than favourable ring for many school leavers. Employment is less secure, and industrial relations, according to Shirai (1986, 87), are often 'premodern'. Only 7% of employees in companies of less than 100 employees are in unions, compared with 60% in companies with 1,000 employees or more in the private sector (1984 figures from Sumiya, 1986, 7).

That does not mean, however, that all smaller companies are exploited by larger ones and use sweated labour, as an extreme dualist view would suggest. Subcontracting accounts for roughly 70% of income of smaller general machinery firms and between 50 and 60% in manufacturing in general. Subcontracting relations, too, have evolved. They were 'market-oriented' on a spot or floating basis before the war, and while the 1950s may have been the golden age of dualism (Nishiguchi, 1989), more recently some subcontractors with a strong technological base have been able to establish greater independence. Controversy exists as to just how much independence there is and how widespread it is, but there is less, it would seem, than in Britain (Sako, 1988).[8]

While subcontracting is common in the electrical, automobile and machinery industries, in steel, shipbuilding and chemicals one finds *shagaiko* (outside-the-company workers) who work at the 'parent' company with parent company tools but are employed by smaller firms. Inagami (1981) contrasts the employment conditions of *shagaiko* and *honko* (regular employees). The former come largely from external labour markets, skilled and unskilled. Small companies try to attract them by offering overtime since the basic rate is lower than that for *honko*. The workers work to make money, although they do help each other out and socialize outside. The security of *honko*, on the other hand, depends on the company, and they accept fierce inter-company competition as their own. They have a sense of community which is aided by expert foremen, grievance handling proce-

Table 2.5 *Employment by company size (Japan, manufacturing)*

Size (employees)	Employment (%)
1–9	12.4
10–29	15.9
30–99	18.6
100–299	14.4
300–499	5.3
500+	32.7
Public, etc.	0.1

Source: 1987 *Rodo hakusho* (White Paper on Labour), appendix 23.

dures, flexible job assignments, planned rotation and so on. Many workers, however, are in a midway position.

The above is illustrative of the variety of employment relations in different companies in different industries and sectors.[9] While the term 'spectrum' economy (Dore, 1984, Clark, 1979) – and labour market – more accurately expresses this variety than 'dualism', workers in smaller firms are generally employed under conditions less favourable than those of larger firms, and for a good number the Berger–Piore description (1980, 16–23) of secondary labour market culture would be reasonably appropriate. In short, there are a variety of employment relations in Japan influenced by legislation, ownership, place in the overall industrial structure, and so on, which have evolved.[10] Smaller firms may seek to emulate the (organization-oriented) employment relations of larger firms, but most do not have the resources to do so, and exhibit employment relations closer to the market-oriented model.

Britain

As in Japan, the public sector in Britain has traditionally been characterized by a high degree of employment security. Unlike Japan, however, the right of most public sector workers to strike and bargain collectively has long been recognized. In recent years, though, public sector unions in Britain have become more militant as a result of incomes policies, imposed ceilings on public spending, attempts to increase accountability and professionalism, and so on (Winchester, 1983, Lloyd and Blackwell, 1983).

Although employment relations in the public sector, particularly the civil service, are nearer to the OER model than are those of many private sector

Table 2.6 *Labour costs by company size (Japan, 5,000+ = 100*

	5,000+	1,000–4,999	300–999	100–299	30–99
Direct labour costs	100	90.8	77.8	69.6	62.1
Monthly payment	100	92.0	81.7	76.2	71.6
Bonuses	100	88.7	67.7	52.2	38.7
Indirect labour costs	100	78.5	61.4	51.6	45.2
Legally required	100	92.1	81.0	73.1	69.6
Voluntary	100	61.1	39.0	30.8	31.1
Retirement	100	73.5	54.4	41.9	27.5
Training/education	100	67.1	44.6	33.1	22.9

Source: Nikkeiren, 1985, 143.

firms, this does not necessarily apply to manual workers, as is seen in their payment:

> As for payment, incremental wages are paid, but not to all: In sharp contrast to manual workers the majority of public-sector employees are paid under some form of incremental payment system, which provides an annual addition to their salary, over and above any increase which may be attained via a national pay settlement. (Elliot and Fallick, 1981, 97)

Manual workers are paid skilled/semi-skilled/unskilled rates which are not incremental (and for a third group, those at the top, discretionary 'salary ranges' replace 'salary scales'). These splits in payment systems reflect splits in most private sector payment systems, while in both sectors in Japan, regular manual workers normally receive incremental pay.

It is perhaps in banking – significantly where there are few manual employees – that we find employment relations closest to the OER pattern. Careful screening of recruits, 'lifetime employment' for core employees, rotation and training, incremental pay and so on have characterized employment. Associations/unions making up the Clearing Banks Union (recently disbanded) have generally been loyal to their employers, who have tried to 'do the right thing' by their staff to maintain loyalty and commitment (Morris, 1986), although BIFU (Banking, Insurance and Finance Union) has been closer to traditional trade unionism. Even in banking, however, there are 'tiers' similar to those described in the railways in Japan, which leaves open the possibility of stratified loyalties which would inhibit the full development of OER and possibly the cooperativeness associated with them.

Table 2.7 *Pay differentials by size of establishment (Japan, 1984 and Britain, 1985; manufacturing)*

Japan		Britain	
Size	Percentage	Size	Percentage
20–99	58.0	1–99	75.9
100–199	63.0	100–199	80.2
200–499	71.4	200–499	84.4
500–999	81.7	500–999	88.8
1,000+	100.0	1,000+	100.0

Sources: Japan: Tsusansho, *Kogyo tokei hyo, kigyo hen*, pp. 8–10 (adapted); Britain: Business Statistics Office, Report on the Census of Production, PA1002.

Minimal involvement, on the other hand, is to be found in industries such as building and construction, shipping and footwear (and dockworking prior to 1967), where there are traditions of casual labour, company sizes are small and labour relations policy is 'contracted out' to employers' associations (Gennard, 1976, Palmer, 1983). In footwear, textiles and clothing, piecework (seen by Wilfred Brown as an abdication of managerial authority and an attempt to regulate worker behaviour by pure pecuniary mechanisms and by Ujihara as the clearest expression of capital–labour opposition) is still common, worker organization is weak and conflict tends to take the form of turnover or absenteeism rather than industrial action (White, 1981, Edwards and Scullion, 1982).

As well as differences between sectors and industries, there are significant differences between firms of different sizes. While 70% of Japanese employees work in organizations of less than 300 employees, in Britain, 70% work in organizations of over 200 (NES, 1986, F100). By establishment size, twice as many Japanese are employed in establishments of less than 100 employees than in Britain. Pay differentials are also less in Britain (table 2.7).

The differences are nonetheless considerable, particularly for manual workers (wages for male manual workers in manufacturing firms of 9 workers or less were 76.2% of those in firms of 200 employees or more, female manual workers 77.5%, compared with 85.8% and 81% for non-manual workers: NES, 1986). Moreover, unions are recognized in far fewer smaller firms than larger ones (Millward and Stevens, 1986, 58), there is less grievance handling machinery, less collective bargaining, and so on.

In contrast to those who emphasize direct control and relative exploitation of workers in smaller firms (e.g. Friedman, 1987), however, others in the Ingham (and Bolton) tradition emphasize greater flexibility in employ-

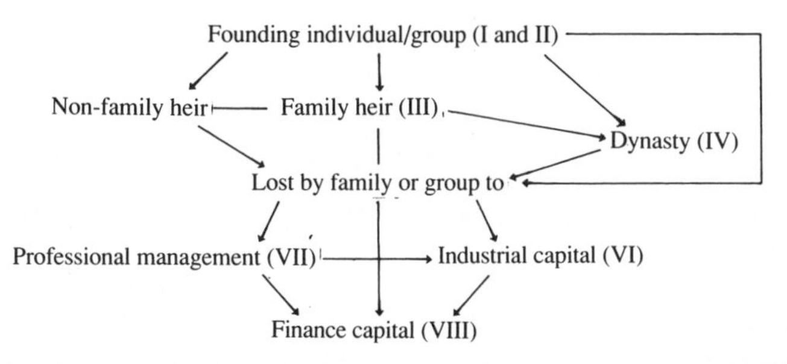

Figure 2.2 Stages of corporate control
Note: The arrows are drawn one way, but this is illustrative only for categories VI, VII and VIII in particular.
Source: Nyman and Silberston, 1978, 89

ment and moral involvement of workers. Asked why they did not go to large firms with higher wages, the majority of respondents in small firms in Ingham's survey (50% of whom had worked in large firms before) replied that they did not like the type of work or that they did not like large, impersonal companies (1970, 96).

The size effect in Britain is less than in Japan in terms of pay and, it would seem, less clear cut in terms of conditions and worker preferences. Further, in government statistics and scholarly publications, less attention is still given to analysis of size differences than to differences of occupation, industry, region, gender, etc. These are all symptomatic, one may argue, of the greater permeation of market-type notions in Britain, and of MER as well, with comparatively less size-sensitivity than OER.

Ownership

Several writers have suggested that a major factor in differences in employment relations between (and among) Japanese and western firms is patterns of ownership, which in Japan are supposed to facilitate long-term planning horizons, less direct influence of shareholders and a blurring of the capital–labour divide; in other words, a different kind of capitalism. As with employment relations, labour markets and intercorporate relations, ownership and capital markets evolve. An evolutionary model is proposed by Nyman and Silberston based on the conception of Francis (see figure 2.2).

Ownership in both countries may have very broadly evolved in this way, as suggested by the figures in table 2.8, but not necessarily in detail, and there are important differences between the two countries.[11] First, one feature of Japanese corporate financing is the 'main bank' system in which companies have a special relationship with a 'main' bank, which is not only

Table 2.8 *Share ownership in Japan and Britain*

	Japan					Britain			
	1950	1960	1970	1980	1986	1963	1969	1975	1981
Individuals	61.3	46.3	39.9	29.2	23.9	54.0	47.4	37.5	28.2
Banks }						1.3	1.7	0.7	0.3
Insurance companies }	12.6	23.1	30.9	37.3	41.7	10.0	12.2	15.9	20.5
Pension funds }						6.4	9.0	16.8	26.7
Investment trusts, unit trusts and other	(11.9)	11.2	2.6	3.2	3.3	12.6	13.0	14.6	10.4
Industrial and other commercial	11.0	17.8	23.1	26.0	24.5	5.1	5.4	3.0	5.1
Public sector	3.1	0.2	0.3	0.2	0.9	1.6	2.6	3.6	3.0
Overseas individuals and corporations		1.4	3.2	4.0	4.7	7.0	6.6	5.6	3.6
Charities						2.1	2.1	2.3	2.2

Sources: Japan: Zenkoku shoken torihikijo kyogikai (various years); Britain: The Stock Exchange, 1987.

the main lender (an average of about 30% of loans for each company comes from its main bank, although the proportion has been dropping recently) but often a main shareholder (although legally restricted to 5%) as well.[12] Main banks maintain close relationships with their clients, but their influence – potential or practised – is generally perceived by managers and employees as supportive rather than threatening. They may accept risks on loans in times of financial crisis of their clients, taking the initiative in coordinating action of both debtors and shareholders. They have an obvious long-term interest in seeing their debts repaid, receiving employee deposits and pension funds, receiving agency fees, and so on.

Secondly, although trust banks and especially life insurance companies are major shareholders, their 'social responsibility' brings them under the same administrative guidance as banks, which regulates their portfolios and discourages them from frequent share selling. In this they are viewed as being different from 'institutional' shareholders in the United States and Britain, although the regulatory framework is now under revision.

Thirdly, industrial shareholding in Japan in the post-war period has actually increased, rising especially after revisions of the Antimonopoly Act in 1949 and 1953. Corporate shareholding is not only of a vertical control type, but also horizontal 'mutual shareholding'. This was fuelled in the late 1960s and early 1970s when, with impending capital liberalization, there was a rush by company managers to find 'safe' shareholders for fear of being bought out by foreign capital and later speculator (*seibi*) groups. The 'safe' shareholders they found included companies with whom their company already had business relations, and other companies in the same 'group'. Thus, companies in the Mitsubishi group own controlling shares of each others' companies, with financial institutions in the group being the largest holders. This is an important factor in the rarity of hostile takeovers in Japan.

As Clark (1979) argues, then, shareholding in Japan is very often the expression of a relationship rather than the relationship itself. Although there is a high turnover of shares on Japanese stock markets (25% on the Tokyo Stock Exchange in 1986), corporate-owned or finance-owned shares are seldom traded. Those buying and selling on the stock markets are normally more interested in capital gains rather than dividends, since the pressure on companies to return high dividends is weak, the rate more or less fixed (e.g. at 20% of par value), and less than returns of other types of investment.[13]

In Britain financial institutions have also increased their ownership of shares – from 30.3% in 1963 to 67.9% in 1981, while that of individuals decreased from 54.0% to 28.2% over the same time. Banks, however, own few shares. A number of observers of ownership in Britain such as Nyman and Silberston (1978) and Scott and Griff (1984) have argued that the growth of finance capital does not necessarily equate with short-term horizons. However, the former do note that 'the forces making for greater profits have been much strengthened' (p. 98) and the latter that the growth in shareholding by financial intermediaries enhanced the availability of share and loan capital for expansion and became an important impetus for the takeover boom of the 1950s and 1960s. Okumura also argues that institutional investors in Britain pursue profits more vigorously (and they do not have the same regulatory restrictions in doing so) as agents rather than as principals (as institutions often behave in Japan); hence these institutional investors should be distinguished from Japanese 'corporate' – including finance – investors. British manufacturing is the most concentrated in the world, according to Scott and Griff, this concentration having risen steadily since World War II. Shareholding by industrial companies reflects control and investment interests, and not those of 'mutual' or 'safe' shareholders that is also common in Japan.

That there is strong shareholder pressure to produce (short-term) profits in the former is not undisputed, but as grumblings from CBI conventions show, such pressure does exist, and managers feel constrained by it.[14] Dore (1987, 146) argues that openness to takeovers makes senior company managers especially sensitive to share prices and 'the bottom line of their next half-yearly results', while the studies of British ownership cited above find little support from Berle and Means and others who argue that a 'managerial revolution' has effectively divorced control from ownership: 'It can plausibly be argued that management control was a short-lived feature of the transition from majority control to control through a constellation of interests, and that it was even more tenuous and short-lived in Britain than in the USA' (Scott and Griff, 1984, 107–8).

As Clark (1979, 98) points out, on the other hand, Japanese directors are commonly viewed as senior managers rather than representatives of the shareholders, even if the law affirms the latter. In his study of ownership and control of Japanese business, Nishiyama goes so far as to contend that Japan is no longer a capitalist society (1984, 124), since its companies are controlled by management workers, based not on ownership but on position and dominance:

> The large businesses in today's Japan are controlled via the position and dominance of management workers. While they still take the form of the joint stock company, this is no longer a *Gesellschaft* composed of stockholders but a *Gemeinschaft* composed of workers. It exists to supply workers with the means of daily living, to satisfy their common interest, and, further, to provide for their common destiny. The management workers, who on the basis of their position and dominance control the businesses, are obviously members of this community. Their management objective is the perpetuation of the firm as a communal body, and the pursuit of profits is merely a means for achieving this objective. (Nishiyama, 1984, 124–5, see also 1975)

While Kuwahara (1987) also argues that larger Japanese firms should be considered employee managed, Okumura (1984, 1988) argues that if Japanese directors do not run companies for shareholders, neither do they run them for employees (or themselves), but for the sake of the 'corporation' itself. Both views, however, posit a restricted influence of shareholders. Senior executives appoint each other, Okumura argues, and the presidents of companies which own each others' shares only interfere in the case of scandal or extremely poor economic performance.

These, briefly, are the grounds for arguing that shareholding patterns in Britain and Japan are qualitatively different – Dore (1985a) uses the terms 'financier-dominated capitalism' and 'producer-dominated capitalism' – and that senior managers in large British firms must be responsive to shareholders while in large Japanese firms shareholders, pressure is not as great. In 1975 less than 10% of Japanese directors had concurrent outside posts compared with about 30% of British directors, while 70% were 'pure' inside directors who had been promoted through the company ranks (Okumura, 1984, 142–3; no corresponding figures were available for Britain). This and a number of other factors may produce a closer identification of interests between senior managers and workers in Japan, as can be seen, for example, when both groups join together to fight takeovers. According to a Nikkeiren survey, one in six Japanese directors have been union executives. Managers' wages are normally pegged to those of their workers, they accept cuts in pay in times of retrenchment along with workers and sometimes in excess of them, and so on.[15]

The identification of senior managers' interests may be reflected in company publications. In the larger British firms in this study, more effort and expense was invested in preparing glossy annual reports for shareholders than employees, more even than for customers in some cases, whereas in the larger Japanese firms the reverse was the case, and shareholder reports were very matter-of-fact.

The argument has been that ownership patterns in large firms in Britain and Japan differ significantly. In Britain the right to control belongs to property-owners, the shareholders, and while managers might be able to act independently in some cases, this does not bring them as close to workers as in Japan, where to a considerable degree the right to control is considered to reside with those with 'property' rights in their jobs *within* the company (even if this is not the case according to the law).

It is interesting to note in this connection the third of the three principles of the Productivity Movement in Japan:

> The fruits of improved productivity must, in correspondence with the condition of the national economy, be distributed fairly among management, labour and the consumer. (Japan Productivity Centre, 1986)

Nowhere are the interests of shareholders mentioned (although the above become 'investors, shareholders, employers, workers and consumers' in the Singaporean adaptation: Taniguchi, 1987).

Ownership patterns are likely to have an impact on the development of employment relations. It is presumably easier to build up a community of interests when the power of shareholders to determine company policies and actions is relatively weak, and when directors are pure insiders subject to the same fate as their employees. The type of ownership described above is, of course, limited to larger, private sector firms. Smaller, owner-managed firms in both countries might be closer in terms of ownership effects, although in reality these cannot be isolated completely from their position in the wider industrial structure on the one hand, and wider notions of employment relations on the other, the effects of which might work in divergent directions.

Culture, compliance and power

We have still not addressed the question of whether the alleged greater cooperativeness of employment relations in Japanese companies can be attributed to greater management resources to induce cooperation, in which case they may coerce workers, or whether workers cooperate out of self-interest, or whether it can be attributed to cultural values which predispose people towards cooperativeness or to be deferential towards the group. It is to this question that we now turn. According to Rueschmeyer:

> The power of employers and different groups of workers, as well as power considerations of the dominant groups responsible to different problem situations, make for more or less clear-cut differences in national patterns of work organizations. The impact of power factors often plays itself out over long periods of time and is in complex ways related with other causal conditions; still, it will not do therefore simply to invoke 'cultural differences' in order to account for national differences in work organization. (1986, 97)

Unfortunately, Rueschmeyer offers just 'a few indications of some contrasts between some countries' rather than a systematic elaboration. Weber's conception of power ('the probability that one actor within a social relationship will be in a position to carry out his will despite resistance', 1968, Vol. 1, 53) can only take us part of the way towards explaining cooperativeness or its absence in employment relations. It may help explain, for instance, how 'present cooperative [Japanese] unions came into existence as a result of the defeats of the unions that were bold enough to carry out long-term strikes against the discharge of their members' (Tokunaga, 1983, 315–16), and also how 'no other working class has achieved the degree of *de facto* workers' control on the factory floor which became characteristic of so many large British factories' through a mixture of formal struggle and informal non-cooperation (Hobsbawm, 1984, 192). The confrontational conception, however, does not fully account for what conflicts do *not* occur, or address such issues as cultural and ideological domination, or control of agenda – 'frontier of discussion' as Fox calls it (1985b, 144).

Martin's (1977) conception of the distribution of control of resources tempered by access to escape as indicating the amount of power involved in a relationship is helpful. Large companies in Japan control substantial resources with which to induce compliance; higher wages, large retirement sums, job security (battered somewhat in shipbuilding and steel), scope for providing a variety of jobs without loss of superior conditions through exit to the external labour market, high social prestige, and so on. This power is magnified where there is a high level of agreement on which resources are valuable (the alleged 'unitary' value system that the EPA, 1986, report refers to). Moreover, the costs of exit for the worker increase year by year.

Workers in large companies, on the other hand, also control greater power resources than workers in smaller companies. To the extent that they gain firm-specific resources that the company needs through long years in the company (e.g. knowledge of specific jobs, company operations and so on) and that there is no escape route for managers (firing and new hiring) they also have resources to create dependence. They are also much more likely to be unionized (unions are, according to Hyman, first and foremost a medium of power; 1975, 64, 189). Again, however, power cannot be discussed merely in terms of resources which may be mobilized for confrontation.

With his concept of disciplinary power, Foucault alludes to a quality rather than just a quantity. This concept is interpreted by Smart as follows:

> Discipline is a technique of power which provides procedures for training or coercing bodies (individual and collective). The instruments through which disciplinary power achieves its hold are hierarchical observations, normalizing judgement and the examination. (Smart, 1985, 85)

Hierarchy renders people visible, making it possible to know and alter them. With this knowledge, disciplinary power seeks to correct non-conformity: 'punishment in a regime of disciplinary power has as its object not expiation or repression, but normalization, and along with surveillance (hierarchical observation) it emerged from the classical age as one of the foremost instruments of the exercise of power' (Smart, 1985, 86). The examination combines the techniques of hierarchical observation and normalizing judgement – it increases visibility, and leads to the collection of individual files which aids the process. These are the tools, then, for a restorative rather than a retributive justice. According to this conception, power relations are based on definitions of and deviations from what is 'normal' rather than being based on the ability to carry out one's will despite resistance, although this may be present in latent form.

Large companies in Japan, with their rigorous screening, constant evaluations of employees (of often visible criteria) and mottoes, exercise extensive 'normalizing judgement' and 'disciplinary power'. This is facilitated by close supervision (see chapter 6).[16] Of course in smaller companies the worker is visible, too, but there is a contrast here with larger British firms where manual workers are not evaluated formally, and supervision is not so close.

Japanese unions, for their part, often carry out surveys of their own members before wage negotiations, measuring satisfaction with wages and different components of wages, views of the union and of management, of overtime and working hours, of working conditions and so on, as well as levels of wages and wage components in comparable companies and their own company's financial position (see chapter 4). These surveys help to back up the demands of union negotiators. Hierarchical observation is thus a power resource for workers as well. Force is sometimes used, but in general the power of persuasion is given preference over the persuasion of power.

Large British companies do not control the same resources as their Japanese counterparts, particularly *vis-à-vis* manual workers, in terms of vastly superior pay, retirement sums, job security and redeployment opportunities. A forty-year-old worker will not forfeit as much from exit as he would in Japan in reduced wages, prestige and loss of seniority for the next job. Moreover, as the Ingham study suggests, there seems to be less agreement as to the relative value of the resources they possess in Britain. One can argue that British companies also exercise less 'disciplinary power', particularly over manual employees, with mostly informal evaluations and examinations, less extensive files, and so on. From the managers' side and the workers' side, 'disciplinary power' as a mode or quality of relations is less developed.

But that is not the whole picture. While employment relations may be based on resource control and embody greater or lesser degrees of 'disciplinary power', they are expressed to a greater or lesser degree through cultural symbols and concepts. Pascale and Athos (1981), and indeed many authors involved in comparative work between Japanese and western countries have commented on the separation of economics and ethics, management and moral law, or man as a productive being from man as a spiritual being in the latter. One consequence is that western managers are much more limited in the extent to which they can mobilize symbols to promote goal convergence, supposing they have the inclination in the first place.[17]

Large Japanese companies, on the other hand, are aided by control of greater resources mentioned above, are more able to mobilize concepts which stress harmony of interests, or serving a higher good:

> Indeed, to judge from the mottoes and philosophies of the large corporations, their goals have become inseparable from those of the nation as a whole. (Werskey, 1986, 10)

Consider the mottoes of one of the larger Japanese factories in this study written at the top of the company newspaper:

> The company is a community. In order to achieve this and contribute to the well-being of society:
> On the basis of agreed cooperation and shared trust we should strengthen solidarity inside the company;
> We should market original and high-quality goods;
> We should seek constant self-improvement and enlarge our occupational skills;
> We should build a bright and peaceful life based on productivity improvement;
> We should try to make a happy workplace, maintaining health, sincerity and truthfulness.

While concepts such as 'community of fate' (*unmei kyodotai*) can be mobilized to foster integration in large companies in Japan (employees in the factory above often mentioned mottoes of the president regarding innovation), contractual relations between powerful and clearly defined groups have traditionally belied such notions in British companies. Smaller Japanese companies, also without the same resources, 'disciplinary power' or ability to mobilize convincing integrated concepts, are less able to elicit normative compliance from their employees, although employees, in turn, are able to exert less power over their employers as well. An episode in one of the small Japanese factories is illustrative. A young worker remarked quite casually in front of the owner that he didn't want to work there, but he had no choice for the time being. The owner commented afterwards that he didn't really mind – they would find a replacement soon enough, and it was better if the youth left sooner than in mid-career, frustrated that their wages couldn't match those of other companies. This would have been unthinkable in the larger companies.

Returning to our earlier question, is the cooperativeness in large private Japanese companies the result of overwhelming resources that managers in these companies can mobilize, and because workers find it in their interests to conform, or is there a set of cultural values which predisposes workers towards being cooperative and accepting authority? Or is this in itself a loaded question; is there not indeed conflict, as Hanami (1979) suggests, and the claim of cooperativeness a ploy to promote what it purports to describe? There is a certain amount of truth to all of these. What is clear is the greater role of normative reference in larger Japanese firms compared with smaller firms and British firms, which we would expect from our discussion of normative compliance as opposed to coercive or remunerative compliance in the last chapter. Large Japanese companies also control an array of (desired) resources, however, which may be used coercively if necessary. Smaller Japanese companies, and many British companies, are unable or unwilling to mobilize integrating concepts to the same extent, and have fewer desired resources with which to create dependence or compliance, hence rely more heavily on remuneration (which may also be inferior as in the case of smaller Japanese firms relative to larger ones), particularly for groups such as manual workers.

2.3 Recent changes

Employment relations undergo change as a result of internal and external pressures. Widespread changes are underway in British industry that may represent a shift towards what has been described as OER. Drastic changes to employment relations in Japan are always seen to be imminent, often towards a market-oriented direction. In the last section we looked at some of the factors promoting OER in Japan and MER in Britain, as well as limitations in extent. The final section in this chapter looks at recent changes and prospects for movement either in the OER or MER direction, beginning with Japan again.

Japan

The 'three pillars' in Japan are continually held to be threatened, by endogenous factors such as changing preferences of younger workers and ageing workforces, exogenous factors such as the appreciation of the yen which has resulted in large increases in investment abroad and a so-called 'hollowing out' of the manufacturing base, as well as decreasing establishment size in manufacturing, industrial restructuring and the rise of tertiary sector employment. 'Headhunting' companies have mushroomed in anticipation of greater labour market mobility.

Actually, the headhunting market is still limited, and the number of 'fixed' employees according to the definition earlier has risen, even in the 1980s: from 42.9% of males in 1964 to 50.9% in 1974 to 55.9% in 1984, and for women from 15.7% to 20.4% to 20.6% (Keizai kikakucho, 1986, 137). The rise in the number of part timers has come about with increased female participation and at the expense of temporary workers rather than those with 'fixed' employment. There are some signs, however, of a rise in mobility of younger workers.

Many people argue that internal career structures are becoming an increasing liability to companies. Ageing workforces not only increase labour costs (0.7% per year, according to one report), but older workers are supposed to be less adaptable to technological change, the pace of which shows signs of increasing if anything.[18] Such considerations are a major factor behind the expansion of 'intermediate' labour markets of companies in the same group or *keiretsu* (Itami and Matsunaga, 1985). The practice of *shukko*, or 'loaning' workers – mainly to other companies in the same group – has grown rapidly, to the point of more than 7% of regular workers according to a recent survey by Inagami (1989, 33). A disproportionate number of these are older workers, who may or may not return to their original company. 'Internal' labour markets represent a compromise between pure internal and external labour markets. Older workers are not (normally) thrown out, thus intermediate labour markets may represent an extension of the 'lifetime employment sphere.' In some cases, particularly department stores, employment is being carried out explicitly on a group basis.

Within the company, promotion bottlenecks caused by ageing workforces can damage morale. 'Specialist' as opposed to managerial posts have been instituted in more than 80% of companies with 1,000 employees or more with the purpose of easing such bottlenecks, as well as increasing efficiency and professionalism, and catering for diversified preferences. Strategic training of specialists is a further impetus to utilizing intermediate or group labour markets.

Changes in wage systems are somewhat more cautious, but most surveys show that companies intend to weight 'ability' more highly in the future, and decrease the influence of *nenko* principles. 'Job-based' wage systems are being experimented with. As Cole (1971, 89) noted, experiments with job-based wages in the 1960s caused numerous difficulties because they clashed with existing practices. They became associated with adversarial industrial relations and inflexibility in western countries. Instead, companies gave greater weight to person-related components such as 'ability'. Present reforms elaborate on these rather than abandoning them or *nenko* principles. In systems like Matsushita's 'new job-based wages' introduced in 1986, a 30% living component is retained, and age is also important in the job component, which bears little resemblance to the market-type wage described in chapter 1. It is worth noting that *nenko* wages have the support of a large proportion of workers even if many are prepared to accept a higher ability component.[19]

Enterprise-based unionism is also supposedly becoming weakened. Organization rates have decreased from 30.5% in 1982 to 27.6% in 1987 (Rodosho, 1988, 188) and are now below 27%. Behind this drop are factors such as decreasing establishment size, the rise in tertiary sector employment, and the inability of unions to organize new workers and industries. Slow economic growth has reduced the room for enterprise unions to manoeuvre (Koshiro, 1986b). Many young workers are not interested in union activities, especially group activities such as rallies. Unions are thus having to adapt their activities, communications organs, slogans and even colours to keep the interest of members and, matching trends in group employment, are coordinating more activities with other unions in the same group, to the extent of joint agreements on some issues (see Inagami and Kawakita eds., 1988, and *Asahi Shinbun*, 7 November, 1987).

In spite of all the arguments, including some rather vociferous ones from the Keizai Doyukai of late calling for the abolition of 'Japanese-style' employment practices, it does not seem as if the present state of employment relations in manufacturing is in for radical change in the immediate future. Rather, modifications will inevitably be made, multitracking will become more widespread with greater differentiation of special groups as well as greater group coordination, but for the moment these will function more to preserve 'Japanese-style' employment relations than to radically alter them. After all, they are given as a major factor in Japan's economic success, and abandoning them altogether would to many managers run the risk of jeopardizing that success.

Britain

In Britain there has been discussion of many of the same issues; decline in the size of manufacturing establishments, organizational problems of trade unions, polarization of the workforce into a core and periphery, and so on. There is also talk about the 'Japanization' of British industry, most noticeable in the automobile industry.[20] Brown is a keen observer of recent changes, and his description sounds very much like a process of increasing differentiation, with one trend – the dominant one – representing a shift in the direction of OER:

> At one extreme it is becoming easier to play the labour markets for some types of work, obtaining relatively cheap and easily disposable employees of the required competence, and allowing their anxiety about job loss to sustain acceptable effort levels. At the other extreme, it is becoming easier to build up a package of employment, training and payment practices that elicit higher labour efficiency through the very different route of cultivating commitment. With the expectation of employment security and with the opportunity to acquire fresh training from the employer as and when required, employees

> are more likely to cooperate with technical innovation, to comply with flexible working, and to bear with tedium. Although the second approach tends to isolate the employee from the external labour market, it is misleading to characterize it as involving an 'internal' labour market. The administrative techniques used to extract increased productivity are the antithesis of market processes; far from simply using a price system to achieve the efficient allocation of labour of discrete skills, they use a variety of devices to motivate labour better to apply and to transmute its skills. (Brown, 1986, 162–3)

Companies are also attempting to heighten 'organization consciousness' through profit sharing and employee share ownership schemes, which have been encouraged by the 1978, 1980 and 1984 Finance Acts. In June, 1986, there were 562 APS (approved profit sharing), 541 SAYE (Save as You Earn) and 1,676 discretionary share option schemes in operation (Smith, 1986). This is still far behind Japan, however, where in 1986, 87.9% of listed companies had employee share ownership schemes (Kuwahara, 1987).

'Harmonization' of conditions of manual and non-manual workers 'might be the most significant development in the coming years' according to Grayson (1984, 22). Although Arthurs argues that 'the need to break down status barriers can be said to have become part of the conventional wisdom of "good employee relations"' (1985, 17), he concludes his study on a sombre note, which later chapters in this book will also echo:

> [Harmonization and single status policies] are unlikely, however, in the absence of other changes, both at work and in the wider society, to alter fundamentally the attitudes of manual workers. Single status and harmonization policies can help to break down barriers at work, but they do not provide easy answers to the 'Us and Them' division in British industry. (Arthurs, 1985, 28)

In the factories in this study, harmonized pension schemes, sickness and holiday schemes were widespread, but 'harmonized' wage systems ranged from individual hourly rates to individual salaries (which, in one case, were clustered around three points representing the old skilled, semi-skilled and unskilled divisions), and in the larger unionized companies, pay systems for manual workers were clearly moving in the direction of having three or four grades only (quite different from Japanese payment systems), which reduced individual incentives but was designed to promote flexibility and remove past problems of fragmented pay structures. Individual incentives were being replaced with company-wide incentives, again, to promote 'organization consciouness' but these schemes were widely opposed by shop-floor workers. While there are signs of British companies moving towards Japanese companies in terms of employment relations, then, some developments are in a quite different direction.[21]

Unions, particularly the 'new realist' unions such as the EETPU, are changing their image in Britain, as in Japan. Brown (1986, 165) argues that 'the structure of trade unionism, originally developed for the strategy of employee solidarity, is increasingly being shaped to the needs of employers.' Certainly, as the BWIR surveys show, the centre of gravity in industrial relations in the private sector has increasingly become the establishment or enterprise, which would facilitate the moulding of industrial relations by individual employers. This does not mean, however, that formal unions are moribund. Seventy-five percent of stewards in the BWIR survey reported meeting a paid official of their union in the previous year, and half reported joint meetings with both their paid officials and management (Millward and Stevens, 1986, 126).

Regarding recent trends in employment relations, Norman Willis of the TUC comments:

> The concept in management circles now is 'human resources management' – seeking to ensure commitment and flexibility from employees in exchange for relatively high pay and job security. This is far too fashionable.

Management are trying to do this without the involvement of unions, he said, but 'just as our ascendancy was exaggerated by our enemies in the 1970s, so is our decline being vastly exaggerated by our enemies today. It is vastly premature to be writing our obituaries.'[22]

While significant changes are underway in British industry, these are being shaped by existing institutions and culture which are the products of long historical development.

3

Employment relations 1

The following two chapters look at employment relations in the 18 factories. This chapter looks at employment and payment, and chapter 4 looks at industrial relations.

Chapter 1 suggested that there would be differences between OER and MER-type factories in the recruiting and subsequent employment of workers. OER managers prefer to recruit new school leavers and train (and socialize) them in the company, are very rigorous in the selection process because they expect their employees to stay long-term, and plan careers for them within the factory or company. The influence of the personnel department and concepts such as 'human resource development' are greater in OER factories.

In OER factories wages or salaries are person-related, reflect the (evaluated) contribution of the employee to overall organizational goals, and reflect the expectation of long-term membership and the need to cultivate loyalty and ability long-term. MER-type payment, on the other hand, is paid for the performance of particular tasks or for particular skills (whether or not they are all being used), the contractual nature of which is not premised on the notion of long-term membership.

The focus of industrial relations in OER is confined largely within the boundaries of the factory, whereas MER industrial relations organization is contoured to reflect the skills or labour power being bought and sold. OER factories place more emphasis on consultative communication, and the notion of 'common destiny' is more readily accepted and acted upon by all parties than in MER. Chapters 3 and 4 will investigate the 18 factories in view of these arguments.

Noticeable in the following account is the greater prevalence of 'norms' in the Japanese factories. While average lengths of employment were not strikingly different between a British and a Japanese factory looked at, for example, 'lifetime employment' was the norm in the latter, but not in the former. Where internal mobility was not so different, rotation would be a norm in the Japanese factories but not in the British. Such norms are a feature of OER, it was argued, but it is important to distinguish between norms and actual practice.

Another feature of both employment and pay systems is inclusiveness; all regular employees were under the same system in most of the Japanese factories, while there were different groups in the British factories such as apprentices and graduate engineers, or manual, clerical and technical workers for pay, despite 'harmonized' pay. Innovations in these areas had been more comprehensive in the larger Japanese factories. They both influenced and were influenced by industrial relations.

3.1 Employment

Recruitment

J45 (Japanese factory of 453 employees) ran its own high school from 1947 until 1969, but closed it when most pupils of the calibre it was seeking started going on to high school and it found the quality of its junior high graduate recruits dropping. Nowadays it sends recruiting advertisements to high schools all over Japan – 3 out of 10 recruits in 1987 were from the northern island of Hokkaido – particularly to technical high schools for manual workers. Those who apply are given an aptitude test which is supposed to measure learning ability and character, weakness under pressure and so on, and which the personnel manager rates highly: 'You can tell a lot about them from the test, how they'll be in 10 years time. We used to have a high turnover of women, but we don't have that now.' He and the general affairs manager carry out subsequent interviews. They recruit about half of the high school graduates they interview and about 20% of the university-graduating mechanical engineers. They are laxer for electrical engineers however. In principle J45 recruits new school and university graduates only, but there were three mid-term recruits in 1986 (see table 3.1) – two electrical engineers and one draughtsman.[1]

All April recruits attend a three-day orientation course in which they listen to speeches from the directors and department managers, who talk about the company, their departments, and matters their departments deal with which are relevant for the recruits. Two days are spent at headquarters, and the third at the factory. The recruits are also taught the rules, safety matters, and are given time to get acquainted with one another. After these three days the female recruits are posted, while the males have three months of training, one month in each of three sections according to a schedule which is laid out quite meticulously by the personnel department. Subsequently they (all) get together for two hours a month for the remainder of

Table 3.1 *Recruitment, f1986*

Factory	Recruits	School leavers	April starters	Factory	Recruits	School leavers	New apprentices
J1	1	0	0	B4	5	0	0
J2*	2	0	2	B8	6	0	0
J4	3	3	3	B11	3	1	1
J9	14	14	14	B12*	3	1	1
J45	20	17	17	B39	15	4	2
J50	7	7	7	B71	30	24	6
J66	55	41	41[a]	B145	25	23	15
J140*	130	130	130	B80*	2	0	0
J180	180	180	180	B120	20	12	12

*Indicates large batch.
[a]Much of the difference between recruits and school leavers in J66 represents 'loan' workers from a major client.

the first year, when personnel department staff come to talk about how the bonus is calculated, personnel changes, and so on. In this first year they are also taken to observe customer factories.

The orientation courses of the other large Japanese factories are slightly longer; one week for J140/180 and J50, and two weeks for J66. The subsequent training period of J66 includes weekends at the Self Defence Forces (males only, to heighten consciousness about 'group rules, moderate behaviour as an employee, character, promptness and health'), *Seinen no Ie* (Youth House, for similar purposes, and 'how to spend a better youth', 'communicating with co-workers', etc.) and a Zen temple (motivation, self-reflection). Sometimes initial training is less thorough than scheduled; when design departments are very busy, for instance, university graduates may be assigned there from the beginning and not move.

While most of the larger factories preferred to recruit technical high school graduates for manual jobs, J140/180, which recruits centrally, prefers graduates from general-course high schools, and to provide all technical training internally. Reasons for this are that such recruits can be more easily 'moulded' to the company image and needs, they are often of a higher calibre scholastically, and the company has its own technical training school. Each year 300 high school recruits (10% of all high school graduate recruits) and 120 junior high school recruits are enrolled in the company school. The latter take a mechanical and general course for three years, while the high school graduates are streamed into a one-year mechanical or electrical course. The top students of the electrical course qualify for a further six months' electronics course. For other high school graduates there is a one week orientation course, another one week of basic training at their workplace, and a further three weeks off-the-job training within the first three years of employment.

In the smaller factories the recruitment and orientation process is much more streamlined. Only J9 gives an aptitude test, while orientation consists of welcoming speeches. The norm of once-a-year recruiting of school leavers is, however, evident at both J9 and J4. Manual recruits for the smaller factories often come from technical high schools where the managing directors have personal contacts; a northern provincial city where its new factory will be built for J4, and the local technical school for J2, where recruits go to school in the evenings.[2] J1 was finding it difficult to recruit workers at all, particularly the younger workers deemed necessary to run CNC machines.

Thus the recruitment of regular workers by the larger factories – less so for the smaller factories – is a careful process, carried out with the expectation that new recruits will stay with the company long term. There is a strong preference for school leavers, except where there are shortages in certain (usually technical or engineering) areas, and recruiting is ideally carried out once a year. This notion seemed peculiar to many British managers regarding manual workers (how would you cover for someone who quit half way through the year, for example?) but less so for university graduates, who were, however, more likely to leave than manual workers.

In the British factories there was one group of manual workers for whom the recruitment process was every bit as rigorous as their Japanese counterparts. These were the apprentices. The manager of the main training centre for B80/120 (each of the major factories in the company also had its own training centre) explained that there were five steps in the selection process. First there was an aptitude test, then a second test, then the first interview with a member of the staff, then a second interview, also with a training manager and line manager. Candidates were also counselled during these interviews – sometimes the youths were not very sure of what they want, and sometimes a more suitable job or type of company could be suggested. Then there was a final 'weeding out'.

'An apprenticeship is not just training for a job, but for a career in the engineering industry [industry, note, not necessarily company]... We have to become a second home for them,' said the manager. Considerable energy is expended in that direction. There are family days to show parents around and display the talents of their children, various extramural activities such as canoeing for charity, and the apprentices had recently been shown around the House of Commons by their local MP.[3]

Other manual recruits, however, were commonly not school leavers, they were not being brought into a 'brotherhood or sisterhood' as the apprentices above were, nor were they selected as rigorously. There was a paper screening and an interview, or as at B8: 'I take them to the shop-floor, give them a drawing, and ask them how they would tackle it. I can usually see how good they are from that.' Whereas apprentices might get a whole day of orientation (as at B145), other manual recruits were 'given the dos and don'ts', and in the larger factories, given a safety handbook and regulations. The criteria of selection were commonly craft or production background, but the managers in the smaller factories were more insistent that they were looking for motivated workers, workers wanting to get ahead. 'Their background doesn't really matter, it's their motivation' (B39); 'We're looking for highly-skilled people, but most of all, people with the right attitude. We ask them what they are trying to do with their life, what they think about doing different jobs, and what they think about training and new technology' (MD, B12).

The managing director of B12, however, was only involved in interviewing recruits with 'promotion prospects' – presumably graduates or staff – while most manual workers were interviewed and hired by managers in whose department they would work. Recruitment of manual workers excluding apprentices was more decentralized in the British factories, indicating that they were being employed more for a specific job than to become a member of the corporate community.

There were signs, however, that greater care was being exercised in recruitment than in the past, as at B80: 'Before anyone who could breathe normally could come in here and get a job. Now we are being more particular.' Nonetheless, there was little indication of a *normative concept* of once-a-year manual recruitment of school leavers for a career in the company, which was widespread in Japan, even if mid-term recruitment was not unusual in practice. The concept of employing manual workers for a career with ongoing training was faintly in evidence, but not systematically developed, and there was certainly no equivalent of visits to Zen temples.

Length of employment

While many of the British factories had carried out large-scale redundancies (see table 1.4) – by and large after considerable efforts to exhaust voluntary avenues, including early retirement schemes – they were making efforts to improve retention rates for certain categories of workers, predictably university graduates and apprentices. The training manager of B80 said that there used to be a tradition of 'sacking' apprentices at the end of their time – 'kicking them out so they could get experience elsewhere.' 'Even now, we don't guarantee an apprentice a job at the end of his training. We try to keep the good ones, of course.' The (best) fruits of training were now to be

Table 3.2 *Average ages and lengths of employment (manual workers)*

Factory	Average entry age	Average age	Average length of employment	Factory	Average entry age	Average age	Average length of employment
J1	24	36	12	B4	27	43	16
J2*	27	35	8	B8	30	40	10
J4	24	32	8	B11	27	42	15
J9	22	29	7	B12*	22	34	12
J45	21	38	17	B39	24	34	10
J50	20	40	20	B71	28	43	15
J66	20	34	14	B145	30	45	15
J140*	19	28	9	B80*	30	50	20
J180	19	28	9	B120	25	40	15

*Indicates large batch factories.

enjoyed by the factory rather than the engineering industry. The industrial relations manager at B71 had spent some time formalizing apprentice training and had managed to raise the retention rate of apprentices during and immediately after their apprenticeship from 25% to almost 100%. Managers were also making an effort to retain other workers who applied themselves – not a new phenomenon, of course – and coupled with greater care during recruiting, this might constitute a move in the direction of fixed employment. This was not a *normative* commitment applied to *all* recruits, however, and retention of graduates was a perennial problem; 20% at B71 after 10 years and 25% at B145, compared with around 90% at J50.[4]

Average length of employment and average entry age figures are given in table 3.2. With the redundancies and reduced opportunities elsewhere, turnover figures (excluding redundancies) were very low in the British factories – around 3% or even less, which was similar to the Japanese factories.[5] Average lengths of employment, however, had been lowered by the redundancies and early retirement schemes in the British factories (the normal retirement age was between 60 and 65). *Personnel policies* in the larger Japanese factories actually limited length of employment there by specifying a mandatory retirement age (*teinen*). *Teinen* is often taken as evidence of 'Japanese-style employment'. It indicates the expectation on the part of the company that workers will stay until retirement, with provisions for regular pay rises until or approaching that date. The government is prodding companies to raise the retirement age to 60 or above with the view to raising the national pension age to 65. The mandatory retirement age for J140/180 was 60, as it was for J66 and J45, 58 for J50 and 55 for J9, although pay increases stopped in all cases after 55. In the smaller factories, loyal

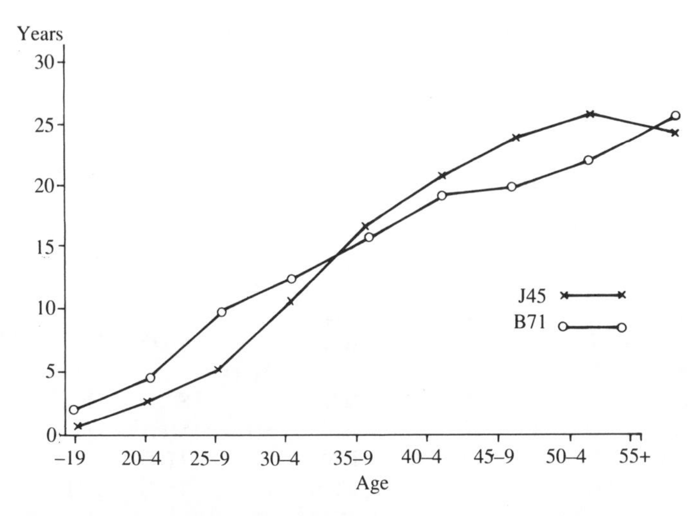

Figure 3.1 Average lengths of employment by age, J45 and B71

workers could keep on working beyond the age of 60 (one was 78 in J2; he had been with the company since 1947).[6]

There is the additional influence on the average length of employment of the average age of the workforce. In factories like J140 and J180 with very young workforces, the average length of employment was naturally short. There is a clear decrease in average entry age with increasing size in the Japanese factories, however, indicating a greater preponderance of school-leaver recruitment and retention. In most cases the average entry ages are considerably below those of the British factories. Both of these could be predicted from earlier discussions.

Nonetheless, a calculation of length of employment by age of workers in B71 (which was located in a rural area, and had many long-serving employees) with J45 (which was not unusual for the Japanese factories and did recruit some mid-term workers) shows a curve which is only slightly flatter for B71 (figure 3.1). According to the definition in chapter 2, 63.6% of B71 employees were under 'fixed employment': 59.2% of manual workers and 73.1% of non-manual workers (many in the non-manual figures, however, were long-serving ex-shop-floor workers, as we shall see); while the corresponding figures for J45 were 79.5%: 76.4% of manual workers and 75.4% of non-manual workers. Again, in the British factories there was no normative concept of 'lifetime employment' just as there was no normative concept of once-a-year recruiting of new school leavers, but there were in fact many long-serving employees whose employment was 'fixed'.

Table 3.3 *Backgrounds and tertiary education levels of selected managers*

Factory	Production manager	Production engineering	Manufacturing manager	Tertiary education	Factory	Production manager	Production engineering	Manufacturing manager	Tertiary education
J1	—	—	S	0%	B4	—	—	S	6%
J2	—	—	S	5	B8	S	—	O	15
J4	S	—	U	12	B11	S	—	U	19
J9	S	S	S	25	B12	—	S(H)	U	10
J45	S	P	P	26	B39	T(H)	—	T(H)	4
J50	P	U	U	20	B71	S	S	U	9
J66	U(S)	U	U	26	B145	S(H)	S(H)	U	10
J140	U&S	U	U	8	B80	S	T(H)	T(H)	7
J180	S	U	S	12	B120	U	T(H)	S(H)	?

Note:
S – shop-floor   O – office
U – university   P – production office/planning
T – technical   (H) – HNC/HND
Tertiary education figures for Britain include HNC/HNDs

Internal mobility

Long-term employment in itself does not constitute a career, at least in the sense of a hierarchy of jobs of progressive difficulty and responsibility. Surprisingly, although most of the Japanese managers did seem committed to developing the abilities of their subordinates (and were evaluated on this), manual workers were *not* as mobile as one might expect. We can get a sense of vertical mobility by looking at the backgrounds of certain managers. Table 3.3 includes both job and educational backgrounds, but in only one case (J66) did a university-trained engineer work for any significant time on the shop-floor, hence the categories are effectively mutually exclusive.

As can be seen from the table, the number of managers with a shop-floor background was in fact higher in the British factories than the Japanese factories. Most of these had done craft apprenticeships, a point which we will return to later in the book.[7]

Part of the difference may be ascribed to a higher proportion of university graduates in the Japanese factories, who occupied many of the positions in indirect departments, thus placing a ceiling on advancement from the shop-floor. This is not the complete explanation, however. Even the high school

Table 3.4 *Ex-shop-floor workers in various departments (percentage)*

Factory	Production engineering	Draughting	Sales	Personnel	Factory	Production engineering	Draughting	Sales	Personnel
J45	10	0	5	0					
J50	15	0	0	0	B71	90	95	75	50
J66	4	2	20	0	B145	80	20	50	50
J140	27	0	5	0	B80	50	20	50	40
J180	28	10	—	0	B120	75	25	5	5

graduates in departments like production engineering and draughting had been posted there from the beginning, and those from the shop-floor were the exceptions. When new employees – with the possible exception of university graduates – were posted to their respective departments, they tended to stay there until they were relatively high, although they might experience a number of different jobs in the department. Sometimes employees in Japanese companies are posted to departments like personnel initially, then given experience in other departments before returning to their original departments (Nihon seisansei honbu, 1986), but there was very little of this, either.

Table 3.4 shows the proportion of workers in selected departments with a shop-floor background. Since B39 had no separate personnel function or production engineering function, it has been omitted. The differences are striking, even when the product is held constant. There was a relatively high proportion of ex-shop-floor workers in production control in the Japanese factories, but even there, the proportion was not as high as in the British factories.[8]

Within the realm of shop-floor jobs, too, workers were often assigned directly to sections like the jig and tool room after the initial three months' training, while such sections in the British factories were manned exclusively by craftsmen who had formerly been operators, or fitter-turners. Nor was there clear evidence of greater mobility between, for example, the machine shop and fitting department in the Japanese factories, or even between the different machining operations. The personnel manager at J66 agreed that there was not very much movement on the shop-floor. 'Rotation is something we have to make much more systematic,' he reflected.

J140/180 is known for its worker training programmes. Even there, however, workers were not moved frequently; high fliers may be moved after two years and 'plodders' after ten (although a training manual stipulates a maximum of eight years at one job). A personnel manager who had just heard his friend the production manager of J140 describe the (lack of) movement between machining and fitting, and even between turning and milling was astonished, and afterwards sighed, 'We tell them rotation, rotation, but the actuality is quite different. We mustn't give up, though.'

These findings are no doubt surprising for those who hold views like the following:

> Compared to the US and Britain there are few specialized production occupations and shopfloor work can be viewed as a temporary, though sometimes lengthy, way station en route to a managerial career. (Jones, 1986, 16)

Jones also makes reference to half-daily rotations.[9] When considering mobility, however, particularly in Japanese factories, one must be mindful of several points. First, normative notions of mobility must be distinguished from actual mobility, and senior and even middle managers, whom researchers most often interview, often express the former. Secondly, interjob mobility must be distinguished from job flexibility or task range – a loose definition of what constitutes a 'job' (see chapter 6). Thirdly, there is a difference between promotion within the shop-floor and promotion from it. There were several ranks below foreman in the Japanese factories (e.g. group leader I, II, and assistant foreman at J50), and only one if any in the British factories (in fact, the leading hand or chargehand position was being abolished in the larger factories). In addition to title ranks, there were also numerous wage grades and levels. Fourthly, drastic structural adjustments requiring cut-backs in a department's labour force are somewhat less likely to be dealt with by redundancy and more likely to be dealt with by 'posting' to another department, which does give rise to mobility, but is not necessarily pre-planned.

To be sure there are regular, planned personnel reshuffles. In a reshuffle in 1986, for example, 86 employees at J66 were either promoted or moved. This affected only three people in the machine shop, however, and four people in assembly (out of about 300).

Mobility, however slow, *was* a normative concept in the Japanese factories, even if disrupted by production needs, while the reverse was the case in the British factories. This was expressed bluntly by a department manager in B145: 'We don't have planned rotation. We don't like window-dressing terminology. We move workers where conditions demand it, and we tell them so.' Indeed, there was no planned rotation, but the interviews indicated that conditions (if not workers) quite often demanded movement.

The personnel function

It might seem from the above that personnel policies were not reflected in actual shop-floor practices in the larger Japanese factories. This is not necessarily so; Japanese line managers seemed to be very conscious of their responsibilities for 'raising' their employees – and were evaluated on this – but their time spans were quite long, and they argued that they had to accommodate 'production requirements'. 'The first thing they impress upon you when you become a manager is that people are the assets [*zaisan*] of the company,' commented one manager in J50. Another at J45 commented:

> I have a number of subordinates. I'm always thinking, how can I raise them up. With CNC, for example, I think who will grow the most by working on it, then I approach that person and ask him. Take the laser machine. There's a lad from my old high school who quit university and came here. He was in assembly for a year, then we decided to get the laser machine, and I decided he would be the right person. He's been there for one year now – he's very good – but now he asks me how long he's going to stay there. He says he wants to go back into assembly. I told him if he didn't change his mind in *two or three years* I'd put him back there. If I take him off now, he won't have mastered it. When he comes to the end of what he can learn on the laser, I'll change him.[10]

Despite the fact that there was a lower percentage of personnel staff from the shop-floor in the Japanese factories, there was little evidence of friction between the personnel and supervisory functions (see also chapters 7 and 8) that was sometimes found in the British factories. The industrial relations manager at B145 ('industrial relations' was preferred to 'personnel', as at B71) for example lamented: 'They [line managers] just don't see the need for training. What you have to do is point out to them the amount of money that might be lost [because of the EITB levy] if they don't.' He thought his department's plans were being compromised at the supervisory level, not just in terms of training, but in communications as well – a big factor in the rejection of the company-wide productivity scheme by shop-floor workers, for example, was the fact that the supervisors sympathized more with the shop-floor than with his department. One of the supervisors in turn commented: 'It [the personnel department] gets bigger every couple of months. There are people there I don't know now. I think they should be there to help us, but sometimes I think it's the other way around.'

The industrial relations or personnel functions did seem to be increasing in importance and sometimes size in the British factories. According to a senior personnel manager at B80, who could have been a management school lecturer, his department had gone from being a separate 'people department' to being 'another department' to being a 'department central to the realization of organizational goals'. There had been a change, he said, from:

industrial relations to human resources management,
people department to consultants,
employees as costs to employees as key strategic investments,
personnel as a separate system to one central for success,
reactive to pro-active,
operational to high-level strategic.

Not all line and production managers and supervisors were as sceptical about the role of the personnel department, or so reticent about training. Said the machine shop manager in B71: 'The old concept where you do your training [apprenticeship] and that's it, you're equipped with all you need, is no longer valid. Products and technology are changing, and the environment, too. There's no point now where you can say that's it!' One person in each department at B71 was responsible for training, and his responsibilities had been enlarged to include adult workers as well as apprentices. (In the industrial relations department, the personnel and training post had been renamed 'employee development' to 'get away from the idea that it was for apprentice and engineer training'.)

Despite the alleged growing importance of the personnel function and growing consciousness of the need for ongoing training, as far as could be ascertained employee files were less extensive in the British factories than in the Japanese, except again for apprentices and graduate engineers. Only in B39, an American-owned company, were workers formally evaluated, while this happened in all the Japanese factories from J4 up.

A recent study suggests that the influence of the personnel department in the Japanese factories has also increased slightly (Nihon seisansei honbu, 1986). Influence is not easy to measure. Sometimes the ratio of personnel staff to total employees is taken as a yardstick, but it was difficult to obtain a meaningful measure of these. At some factories personnel staff were involved in wider company activities, and at some headquarters the responsibilities of personnel staff included affairs in the various factories. Roughly speaking, the ratios in the Japanese factories were greater than 1:100, whereas they were somewhat less than 1:100 in the British factories.

Looking at access to top positions, of the five company presidents at J140/180, one had been from personnel. The present president has an engineering background, but the personnel department is said to wield quite a strong influence because it puts forward promotion plans to the board of directors. At both J66 and J45, sons of former presidents have taken over as president, being groomed in various areas such as sales and planning, but with little experience in personnel. At J50, the present president is from sales, although a previous one was from general affairs (which encompasses personnel). Of the British factories, perhaps the personnel director of B80/120 wielded the most influence in his company. This was the company most directly facing foreign – especially Japanese – competition, and where 'human resource management' was most talked about. (There was even talk of changing the title Personnel Director to Human Resources Director). In B145 and B71, however, the personnel manager did not have director status, and in none of the factories had the Chairman or MD come from Personnel or Industrial Relations.

Thus there was some evidence that the personnel department was a more likely springboard into top management positions and perhaps wielded more influence at the board level in the Japanese firms, but there were also differences within the countries. On the whole, the role and influence of the personnel function in the Japanese factories appeared to be stable, and on a higher and more central plateau than that yet attained in the British factories.

To summarize briefly, while much care was taken in recruiting apprentices and graduate employees in the British factories, this was extended to all regular recruits in the larger Japanese factories, although following orientation, much more training was given to male employees than female employees.[11] Once-a-year recruitment of school leavers was a goal, or normative concept, in the Japanese factories, but not in the British factories. Orientation itself in the larger Japanese factories represented an attempt to welcome all new employees into a community, and to give an overview of that community and its role in modern society, whereas in the larger British factories (where it was carried out) it consisted mainly of 'giving the dos and don'ts' and dealing with administrative affairs. Apprentices were trained and socialized to identify with their occupation, the engineering industry and also their company, and once they had served their time, they were considered able to 'stand on their own two feet'.[12]

Turnover was low in both countries, and increasing efforts were being made to retain such workers as apprentices and graduates in the British factories. Retention rates of the latter, however, were much lower than in the Japanese factories. When it came to mobility, quite a lot of movement from the shop-floor – and within it – took place in the British factories, while movement in the Japanese factories was not as systematic or frequent (on the shop-floor at least) as personnel policies would lead one to expect. These findings run contrary to popular conceptions of mobility, and happened despite the professed commitment of many Japanese line managers to training their subordinates, and the assertion of some British managers that mobility was dependent upon production needs. Finally, there did seem to be evidence that the personal function carried more weight in the larger

Japanese factories, and it was perhaps more successful in diffusing 'human resource' concepts to production functions, even if this was not reflected in mobility patterns.

3.2 Payment systems

The differences between the two countries in terms of payment systems were not just quantitative, but qualitative, and it is important to consider these in some detail, given the importance of payment systems to employment relations. I shall therefore discuss in some detail first the payment system in one Japanese factory, and afterwards note deviations in the others, and then follow the same procedure for the British factories. Different concepts of fairness are incorporated into both systems, for managers and workers (not necessarily of equal proportion) as suggested in chapter 2, and both have evolved in the overall context of employment relations, as we shall see. Finally, I will look at the differences between pay for manual machine operators and CNC operators.

J45

J45 was a late developer of sorts. The new payment system it adopted in 1973 was introduced in other large factories earlier in the post-war period. It was the culmination of a move away from piecework for production workers, and of putting manual workers and staff workers on exactly the same basis. Piecework was phased out because of familiar problems; disputes over rates, waiting, scrap and so on.[13] The union was formed over the issue of lowering piecework rates in 1954, an issue which arose again several times in the next few years. Although the rates were still high even compared to the wages of engineers, the union opted for stability in 1962 with hourly rates and a group-based premium for work done in excess of 'standard hours'. The engineers objected to manual workers only receiving this premium, and by 1964 it was applied to all workers.

Prior to 1962 there was a very clear blue collar/white collar divide. If a blue collar worker were promoted to white collar status, he had to officially leave the company and then rejoin. Abandoning piecework, according to the personnel manager, had a strangely stabilizing effect, and was important in diminishing status differences, although attendance did not change noticeably.

The new system adopted in 1973 was both an extension of past developments and something quite new. The explanation in the company newspaper (*shanaiho*) in 1973 began with the following comments:

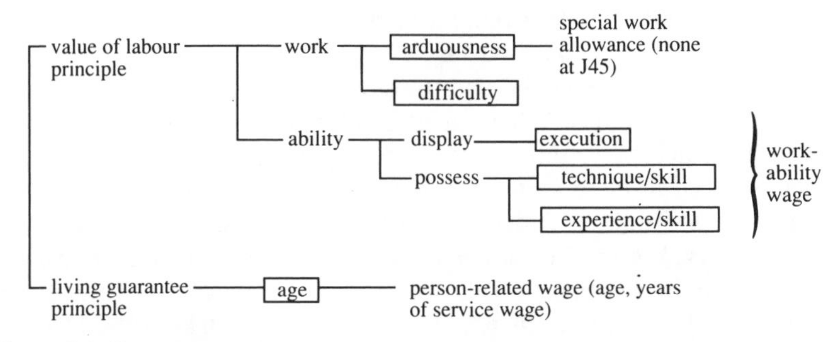

Figure 3.2 Pay system concept, J45

> Wages are what is paid to results achieved through work. Consequently, wages and company performance are not unrelated. Company performance is the result of administrative effort and improvements in skill and techniques of all the employees. Based on this principle, it is possible to say that wages which are distributed on the basis of improvements in skills and techniques, and the use of these, are the fairest. However, since wages form the basis of living, wages should not be so low as to make possible only a low standard of living relative to others, hence a minimum guaranteed wage is provided so that even the worker with the worst [wage] conditions can maintain a living.

Notice the emphasis on 'livelihood' and 'skill improvement', or both needs and abilities. This is restated later: there are two basic reasons why people work, to earn their daily bread and to develop themselves, and if only one of these is reflected in the wage structure, that structure will ultimately fail. The two aspects were represented diagrammatically (figure 3.2).

The pay system adopted has three main components; the age and years of service component, the evaluated work ability component, and various 'allowances' such as a housing allowance, family allowance, etc., as well as a twice-yearly bonus. The 1987 Employment Regulations of the company stipulated that the age component was to be ¥59,000 for an 18-year-old, rising by ¥430 each year to the age of 55 (¥75,000), thence level until the retirement age of 58. The years of service component was to start at 0 and increase by roughly ¥2,500 each year, although the amount of yearly increase varied slightly, peaking at over ¥3,000 in the early 30s.

Pay-related evaluations are carried out twice a year, with a third for promotion considerations. Workers are evaluated according to ability, achievement and attitude. Ability for manual workers is judged according to knowledge of work, skill, mastery of work and judgement (for office and technical workers the second and third criteria are replaced by execution and negotiating ability). Achievement is judged according to the amount of work done, accuracy, safety and reporting and communication (safety being replaced by methods improvement in the case of office and technical workers). Attitude is judged according to 'positiveness', sense of responsibility, keeping of rules and cooperativeness (same for clerical and technical employees; there are also categories of managers, administrators and specialists). Points are given, summed and placed in an A B C D E type distribution, which affects the rate of rise of the evaluated part of the salary. Achievement and attitude are given more weight for bonuses, and ability for wage rises.

Based on years of service and the evaluations, employees are slotted into grades and steps within those grades. There were ten grades and seventy steps within the grades in 1987, the amount for each also being displayed in the Employment Regulations. Promotion is automatic up to grade 3. Grade 4 is the foreman rank. Those at the top of grade 3, however, earn much more than those at the bottom of grade 4; grade 3 starts at *c.* ¥65,000 and ends at ¥140,000, and grade 4 starts at ¥85,000 and ends at ¥175,000. Promotion in the lower grades, then, takes place with the fulfilment of age and years of service requirements, but is faster for those with high evaluations. Thus it rewards plodders and high fliers differentially, but it provides an incentive to both. High school graduates, for example, may be promoted to grade 2 after 4 years if they are very quick (in which case they would be on roughly the same rate as university graduates, who start at grade 2), or 7 years if slow, and from grade 2 to grade 3 in 5 to 12 years. Fast university graduates may move from grade 2 to grade 3 in 4 years (see figure 3.3).

In devising the pay system, senior workers were enlisted along with managers to classify and describe jobs, the skills that were needed to perform them, and the criteria for the evaluation of those jobs. Upon the basis of these, training and personnel activities would be carried out. The same company newspaper stated:

> According to these external and internal conditions, there is a need for a clear standard of evaluation, and the need to show concretely how the individual can develop through work.

The aim of the evaluation system was not only to reward or punish performance, but to spell out clear steps of 'self-development' (which would enable the worker to earn higher wages and promotion). Providing clear descriptions and criteria was also intended to facilitate movement between job types.

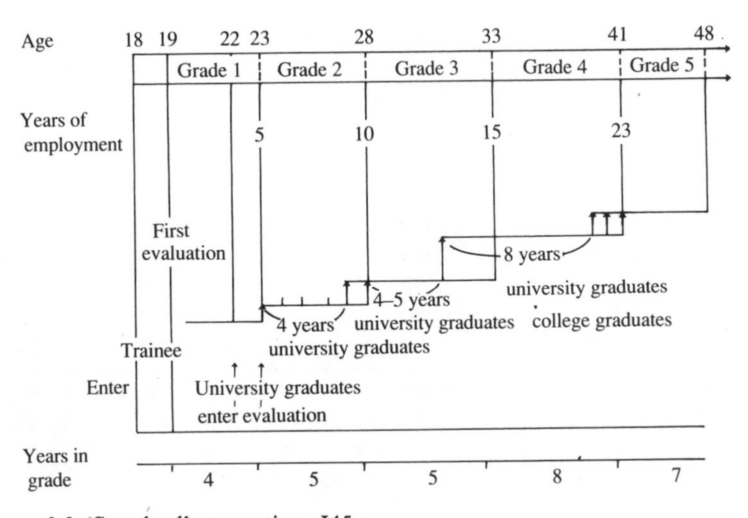

Figure 3.3 'Standard' promotion, J45
Note: After age 33, promotion times for university graduates and college graduates differ.

There is one more portion of the salary which varies and is designated an allowance: a premium (*shoreikyu*) for exceeding standard hours and sales targets, applied across the board, and averaging ¥4,000 in 1979 to a high ¥12,000+ in 1981 and 1985. Other allowances include a family allowance, housing allowance, separation allowance (where the worker has to work away from family), attendance allowance, commuting allowance, and 'post' allowance (e.g. ¥3,000 for group leader – *kumicho*; ¥6,000 for foreman or chief clerk – *kakaricho*).

In terms of a minimum wage, the company promised to pay at least 75% of 'standard living expenses' based on a model of marriage at 27, third member of the family at 30, fourth at 35 and fifth – a parent – at 40, calculated from figures for males used by the National Personnel Authority (Jinjiin) and Prime Minister's Office surveys. (This 75% excluded allowances, overtime and presumably bonuses.)

The bonus consists of a base rate portion (50%) and an evaluated portion (50%), and is multiplied by the attendance rate. It is in part related to company performance, but in actuality does not vary as widely. When sales profits slumped by more than 50% between 1978 and 1979, for instance, the bonus fell by only 5%. The overall average for bonuses, expressed as a multiple of basic monthly salaries, is the subject of collective bargaining in the autumn and spring, and it takes a serious drop in company performance to cause this multiple to fall. The average combined bonus for 1977 was 5.04

Table 3.5 *Wage components, J45 (model, ¥, 1986–7)*

Age	18 (school leaver)	35	50
Component			
1 Age	59,000	66,000	73,000
2 Years of service	0	47,000	69,000
3 Work ability	46,000[a]	114,000	189,000
4 Premium (company)	8,000	8,000	8,000
5 Attendance	500	500	500
6 Housing	1,000	2,000	2,000
7 Family	0	16,000	16,000
Monthly salary	114,500	253,500	357,500
Bonus (1986 total)	386,000	1,275,000	1,738,000

[a]A starting component in lieu of evaluation for the first year. Figures in ¥.

months' average basic salary, 5.07 months' the following year, then 4.62, 4.69, 4.81, 5.21, 5.10, 5.19 and 5.31 in 1985.

We can get an idea of the spread of wages (omitting overtime) by age from figure 3.4. Note, however, that managers' wages are not shown. The highest paid would earn almost ¥6.5m per year at 40 and about ¥9.3m at 56, which is two-thirds more than the lowest-paid worker at the same age, and about 5.4 times that of a new 18-year-old worker (the president (*shacho*) would get about 7 times as much).

As for indirect labour costs, in 1985, the company paid the equivalent of 8.9% of direct labour costs in statutory and 4.2% in non-statutory welfare benefits, not including retirement money (6.2%) and training (0.5%). Of the 1.4% of workers' earnings (excluding bonus) which must be paid for 'employment insurance', the company paid slightly more than 60% (minimum legal requirement is 50%); of the 7.9% for health insurance it paid 53% (minimum requirement 50%); it paid half the 6.2% pension deduction (50%), and all of the 0.8% accident insurance (100%).

The other Japanese factories

Despite J45's intention of using job descriptions and evaluations as the basis of both pay and employee development, the employee development link has in fact not been systematically developed. Workers are only interviewed when there is a considerable change in their evaluation, or when they are dissatisfied with their evaluation. The link has been more explicitly devel-

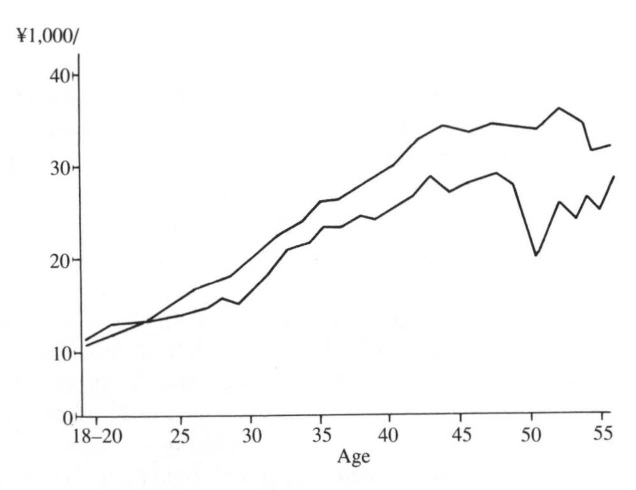

Figure 3.4 Age–earnings of union members, J45

Note: Figures are in ¥10,000 and are for monthly payment without overtime or bonuses. The top line represents the maximum and the bottom line represents the minimum. Low dips (in the minimum) show mid-term entry.

oped in J140/180, where workers not only have to fill in a self-evaluation sheet, but they fill it in with respect to the achievement of goals they have already set for themselves in consultation with their supervisor based on the job descriptions – a kind of MBO (management by objectives) system for all employees. They can also see the criteria by which their supervisor evaluates them since that sheet is on the back of the sheet they fill in themselves.

As in J45, a union book at J140/180 shows that the 'life cycle' or 'living wages' concept is very important for workers. Precise calculations are made as to how much money a worker has to save in order to be able to put down one third deposit on a ¥22.5m home at the age of 32, how much is needed for the education of children at various ages, and so on. This information is used for advising members on savings plans and loans, as well as in collective bargaining.

The pay systems of the other large factories varied in detail, such as more or fewer grades, flat or variable increases for years of service, and so on, but not greatly from what was described for J45. Negotiations were in fact carried out on the basis of what other major companies, not necessarily in the same sector, were paying their employees for each of the different wage components.

There was more variation in the smaller factories. By and large, pay was decided unilaterally. The same names for components of the pay were used, but the 'evaluated' parts were not systematically elaborated, nor were the criteria available for the scrutiny of the employees. They were sometimes used to 'balance out' the pay, so that in the end, the pay in the smallest factories resembled more a market wage than an organization wage, although a market wage related to the type of person the management wanted to attract and how badly they wanted to keep that person.

J2, for example, paid 'somewhat more' than the local rate (a rather isolated area with few other firms of any size nearby) to keep its workers happy. There was both an 'evaluated' component and a 'special allowance' which were largely differences on paper; they were balanced out, and if it were felt that a younger worker were not receiving enough, his housing allowance might be 'topped up'. The owners were concerned to keep the base wage low because that was the basis of calculating retirement and severance money. Younger workers still at school were paid an hourly rate. This was decided after consulting the school teachers in order to inculcate steady working habits.

While J9 still had a *nikkyu gekkyu* (day wages on a monthly basis, with reductions for absence) system for all employees below the position of section manager, the managing director at J4 (an ex-IBM employee) had moved on to a full salary system for all regular employees in 1980, and was introducing an evaluation system designed in consultation with senior workers.

Overtime was an important element in managers' and workers' comments about pay in the smaller factories. One worker in J2 who had also worked at a larger company sighed:

> In small companies you have to work long hours. Man shouldn't work such long hours – it's not natural. . . I do 50–60 hours overtime a month. I have to do that to make the wages I got in my last job, and the bonus is only half as much.

Indeed, with his overtime he was earning the same as someone the same age (36) in J45 without overtime, and his bonus was half as much. There was little difference between the salaries of school leavers in J2 and J45. In fact, the former were paid 10% more, but the salaries of the older workers levelled out around 40 at about two-thirds more than a school leaver, whereas at J45 (and the other factories from J4 up) they kept rising. The influence of the pay systems of the larger factories can be seen in the terms used by the smaller factories, and by the fact that J4 and J9 were striving to create similar systems, while a predominant concern for the going rate for certain types of workers was only evident in the smallest factories.

Table 3.6 *Wages of direct operators, B71 (£, 1987)*

Trade	National minimum	Fallback rate	Standard 39 hours
Pattern checker	102	—	125
Patternmaker, moulder and coremaker (skilled)	102	—	125
Numerically controlled lathe setter	102	115	124
Fitter, turner, welder, plater, borer	102	114	123
Driller I, grinder, capstan operator, general operative	89	107	116
Driller II, grinder, fettler, sawyer	88	106	115
Machine and fitting shops direct helper, trimmer stamper	85	104	112
Postal packer	73	102	109

Note: Other grades omitted. Standard 39 hours for a labourer = £107.

B71

The pay systems at B71 had evolved, too, and were in the process of further evolution at the time of the interviews. Many of the incentives to change were similar to those of J45 – problems associated with piecework, company attempts to heighten 'company consciousness', harmonization, flexibility promotion and so on – but the path taken has been different.

B71 abolished piecework in 1968, six years after J45, to put a stop to haggling and 'working off a fag packet' by rate fixers, as well as to combat creeping wage inflation. Consultants were called in, and at that time the production engineering department and work study group were established. Jobs were evaluated, and from these individual incentives with a fallback rate were established. This was done in collaboration with the unions, and recognized existing differentials.

The result was 25 grades of hourly paid workers, grouped as direct operators, indirect operators and operators in indirect departments. The highest grade now is that of indirect chargehand in indirect department, closely followed by those with similar jobs in direct departments. CNC operators are not far behind, and so on down to labourers. For direct operators, the (abbreviated) picture in 1987 is shown in table 3.6. These figures do not represent what was actually earned, however. The fallback rate was roughly 93% of the standard rate, but in fact the individual incentive portion had crept up to about 25% of the standard week (it reached 33% at one time), and the top operators were being paid £160–5 per week in 1987. While the original intention may have been to pay a rate for a job, it now paid the person, since it was given whether the person was doing the particular job or not. There is still reference to a market 'rate' however.

The age-earnings profile of hourly workers is shown in figure 3.5, and is compared with those for non-manual workers, and manual and non-manual workers at J45.[14]

The pay system at B71 was ágain in the throes of change, however. The change was intended to promote flexibility, as well as to balance creeping incentive payments, which had caused problems in differentials. Electricians claimed their wages had become incompatible with those of the operators, and there were problems between hourly paid and staff employees in general. When the 'imbalance' (from the staff perspective) was rectified, the hourly paid workers started an overtime ban, and so on.

The 25 grades were being replaced by 4 grades, the 7 grades for fully-skilled workers by one grade. Individual bonuses were being replaced by a company-wide productivity scheme based on saleable standard hours divided by the number of attendance hours (of direct workers, and the number of indirect workers needed for the production).[15] Said the production manager: 'There will be no incentive for waiting for scrap, and an incentive for all in the firm to ensure that the greatest number of standard hours are produced.'

The proposed changes were put to the workers and supported by a ratio of 4 to 3, but predictably opposed by the top earners: 'The new wage system? It's ridiculous. It means a skilled man doing nothing will get the same as me. I've worked hard and I've got responsibility. I might as well just go and sit on my machine'; 'I voted against it. The young apprentices who didn't know voted for it and so did the older workers.' While the managers hoped for more flexibility and cooperation, the changes went against the notions of individual contract of the highest-paid manual workers.

The clerical and technical workers as well as managers are all on separate schemes. There are 5 clerical grades, 4 technical ones, and the managerial scheme is based on the Hay system of grades and evaluations. In 1976 the new owners of the company (it has since been acquired again) requested a yearly evaluation for all staff members, which is carried out in October to separate it from the wage negotiations in April. The evaluation system is still used, but 'has been eroded by events over time, as all pay schemes are'. It was claimed that there had been an element of seniority in the pay of non-manual workers in the 1970s, but that had also been eroded away (by mutual consent) during the times of incomes policy.

The principal differences between conditions for manual and non-manual workers, according to the industrial relations manager, was that hourly

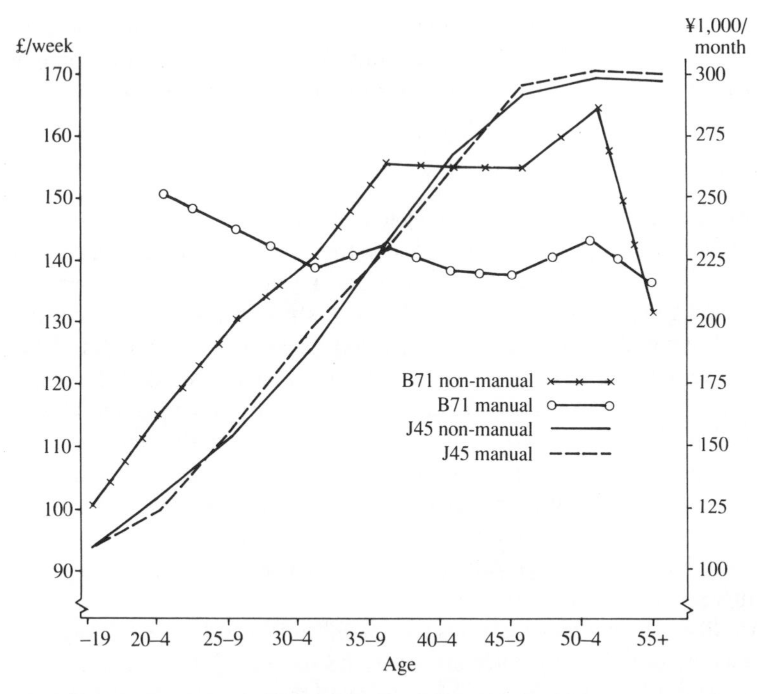

Figure 3.5 Age–earnings profiles at B71 and J45 (manual and non-manual)
Note: All curves are based on actual earnings, not those of 'standard' or 'model' workers.

paid workers had to clock on, and worked 39 hours instead of 37 hours. He was tackling the problem of direct credit transfer for manual workers (some of whom had never had a bank account), but it was pointed out that 20 or so of the staff also took their pay in cash. Holiday and sickness terms had been harmonized, and the pension scheme (which was contracted out) for manual workers was if anything better than that for the staff because it was newer, he claimed. Manual workers paid 3.75% of their earnings for the pension scheme and the company paid 8.2%, while the staff paid 4.5–5% and the company 9%. (The company paid on average 10% of gross wages for national health insurance, as laid down by law.)[16]

The age-earnings profiles in figure 3.5 suggest that the differences went deeper, representing two different types of employment, one calculated to employ and reward employees relatively in the long term and the other not. Moreover, it seemed unlikely that complete harmonization would be carried out if it involved evaluations. The notion of evaluating manual workers sounded preposterous to the industrial relations manager: 'For one thing, evaluations are subjective, and imagine administering the system across 700 workers. There wouldn't be any economic benefits!' He did not explain why it was any less subjective or uneconomic for non-manual workers.

The other British factories

The larger factories were all moving in the direction of fewer manual grades. In 1984, B120 had 120 different levels for 750 hourly paid workers (on 3 sites). 'How could you say to someone getting £120.10 that they were worth less than someone getting £120.12?' (work study manager). The confusion had grown with years of ad-hocery, and the system was replaced in 1985 by five grades of skilled (time-served) workers, three of semi-skilled and one of unskilled. The aim was to reduce the number of skilled/semi-skilled/unskilled grades to 3:2:1 and eventually to 1:1:1. Individual incentives had also been replaced by a company-wide bonus scheme. Managers held that the removal of a lot of other problems more than compensated for the loss of individual incentives. B80 was moving in the same direction, while B145 already had only six grades for manual workers, and a company-wide bonus scheme.

In all of the smaller factories except B8 but including B39, wages were ostensibly individually determined. At B8 for each job there was a beginning, a middle and a top rate. Most were on the top rate. Besides this, there was a 1.5% raise after seven years of service. The switch to individual determination in the early 1980s at B11 had resulted in an increased spread in wages, although there did seem to be a ceiling on how much semi-skilled operators could earn, while at B39, some workers thought that wage differences – from a merit rating of up to 15% of average wages – were being compressed. That merit rating was based on a twice-yearly evaluation carried out by the supervisors according to such criteria as cheerfulness, cooperativeness, punctuality, speed of work, and so on. 'They know exactly what we think of them by their wage,' said the production manager. That would imply, of course, that they compared their wages with those of their co-workers. Some obviously did; the CNC operators, for example, felt that their wages had actually been brought closer together in order to prevent arguments and resentments. They thought the same applied for other groups, too, but were not sure because the system was 'secretive'.

Amongst the smaller British firms, B11 had gone furthest towards removing status differences with its 'single status' scheme, which included the same number of hours of work for manual and non-manual workers. This scheme, according to the foreman, had gone a long way towards eliminating the 'Us–Them' divide.[17] The structural restraints to harmonization were, of course, smaller in the smaller factories. Of the larger factories, B145 had perhaps gone furthest towards harmonization. Manual workers are now called works staff, are paid by weekly credit transfer, are not automatically docked for being late, and are included in the self-certification sickness scheme. Clocking on takes place for all employees earning below £10,800, and it seems to be a matter of time before the hours worked are the same.

CNC and pay

One issue not touched upon so far is the rate of CNC operators compared with manual machine operators. There was no differential in any of the Japanese factories, although at J1 the son of the owner said he would like to have young workers operating several CNC machines and pay them a higher wage, but he would have to do that in a different location or there would be trouble from older conventional machine tool operators. The pay systems at the Japanese factories meant that while moving a worker from conventional machines to CNC or vice versa might involve considerations such as morale, it would not involve considerations of pay differentials.

This was not the case in several of the British factories, however. B71, for example, had a differential of £1.11 per week, the establishment of which, apparently, had been of prime concern to the AEU members. While it may have been substantial in 1972–3 when it was established, it had not risen over the years, and it was paid not only to those working on CNC at a particular time, but to those capable of operating CNC machines, with the result that a large proportion of the craftsmen operators received the premium. A premium of £2 was also given to operators for correcting CNC tapes after they protested that their efforts in that quarter were not being given due recognition. The traditionally more militant workers at B145, on the other hand, never pushed for any differential for CNC work. Their policy, according to the convenor, was that any benefits accruing from new technology should be shared by all workers rather than a select few.

B120 had bought out a differential it was forced to give, but CNC operators were in the top category ('sometimes we inflated the rate to buy change'), while at B80 the top rate given to setters was for setting CNC machines because the 'CNC setters are the best anyway' (personnel manager).[18] The same applied to the operators (there was a £10–12 gap between the bottom of the setters and the top of the operators). At B39, B12 and B11 CNC operators were the highest paid, allegedly because they were the top operators, and not because a higher rate was paid for CNC work, while at B8 CNC operators were given 5–7% more than manual operators. At B4 no special rate was given for CNC work.

In some cases, then, CNC work was explicitly recognized by a higher rate. At least at B71, this rate was given to avoid friction over CNC introduction, but at others managers claimed that they wanted the best operators on CNC, so they gave the highest rate. In the smaller factories where pay was unilaterally decided, it was claimed that the higher wages of CNC operators merely indicated that they were the best anyway. In no case, however, were CNC operators paid less than their counterparts on conventional machines.

We have seen that pay in Japanese factories was largely person-related, varying according to age, years of service, ability, achievement and attitude towards work regardless of what that work was, and various allowances such as attendance and family. While the determination of each of these was laid out in the employment regulations of the larger factories and was intended to shape worker habits and attitudes, in the smallest factories the different components were juggled so that the end result was in effect a market wage for that kind of worker in the local labour market. The larger British factories were moving towards having a limited number of grades for manual workers. This meant that the 'rate-for-the-job' principle had become quite attenuated, although the ultimate reference was still the type of labour that the worker was supposed to be able to supply rather than the worker himself. In the smaller factories, too, managers were very aware of the need to pay competitive rates or risk losing their better workers; the individual determination allowed them to fix their rates more precisely, although they were also using it to provide individual incentives and encourage workers to respond to company objectives. Differences in pay between CNC operators and manual machine operators gave some indication of market orientation, but there were other principles involved as well, such as union concern that benefits from new technology be distributed to all at B145, and the claim in the smaller British factories that the best operators were on CNC, anyway.

It seemed that while pay in the large Japanese factories clearly reinforced the norm of long-term employment for all members of the corporate community and convergence of individual and corporate interests, recent changes reducing the number of manual grades in the large British factories had increased the ambiguity for manual workers as to just what was being paid, and the productivity bonuses were viewed sceptically rather than increasing identification with the company as intended. Following the section on employment, a natural question to ask is whether or not the streamlined grading system was conceived of as a promotion hierarchy. At B145, where some 'semi-skilled' machinists had been upgraded to the 'skilled' category, the convenor did like to see them as such, with new entrants restricted to the lowest grade 5 and promoted internally. The industrial relations manager, however, did not see them that way. Internal promotion would occur *if* (not *when*) the workers had the right skills and there was an opening for them. Changes in the grading systems in the larger factories were to ameliorate anomalies that had crept into pay and to promote flexibility, but not to create an internal career structure or to 'show concretely how the individual can develop through work' and be rewarded for it as at J45. In this the changes in payment were not synchronized with the nascent changes described for employment as they were in most of the Japanese factories.

Regarding harmonization, no distinction was made between manual and non-manual workers as far as the calculation of pay was concerned in the Japanese factories, although different criteria were given different weights in the evaluations. Most of the other distinctions had also been removed, although there were still differences; manual workers would not rise as high as university graduates, at J45 they still took most of their holidays, and sometimes distinguishing terms were used (*genba* versus *stafu*, etc.).

While obvious distinctions were being removed in the British factories, basic differences remained, as the age-earnings profiles at B71 suggested. There were still separate pay categories of manual, technical and clerical workers with different principles of calculation, related mainly through occupational relativities. And even where these differences did not exist in the smaller factories, the pay system itself did not foster a convergence of interest between the two groups, as we shall now see.

4

Employment relations 2

The third pillar of employment relations was influenced by and in turn influenced the first two pillars. The discussion of employment relations continues with a look at the organization, funding and 'mood' of industrial relations in the respective factories. Although there were many features which promoted cooperation in the larger Japanese factories, industrial relations had in the past been quite stormy. There had been a 'stabilization' in the British factories, too, but there were important differences.

The overall discussion is summarized in the final section in which an attempt is made to evaluate the factories on fifteen criteria, five from each of employment, payment and industrial relations to achieve an overall measure on the OER–MER continuum.

4.1 Industrial relations

Organization

All the large Japanese factories had single unions except J50, where a second union was formed following a protracted pay dispute in 1975, while in the large British factories several unions were recognized. This lent a simplicity to the union organization and industrial relations machinery in the Japanese factories and a complexity in the British factories. (The former was more complex, however, in that the same organization served more functions than in the British factories.) Single union organization promoted harmonization in employment and payment conditions for manual and non-manual members – and was in turn promoted by them – although the potential for conflict of interests between these groups was recognized.

All *regular* employees in J45 up to the rank of section chief, including graduates destined for managerial careers but excluding employees in the personnel section, are members of the (enterprise) union.[1] One representative for every 15–20 members is elected every year (table 4.1), although once elected, the representatives tend to serve for 3–4 years. Although there is a

Table 4.1 *Formal industrial relations organization*

Factory	Number of unions	Workers per rep.	Full-time officials	Works Councils/ joint consultation	Levels of joint consultation
J1	0	—	—	No	—
J2	0	—	—	No	—
J4	(1)	13	0	(see union)	1
J9	0	—	—	No	—
J45	1	15–20	0	Yes	2
J50	2	20	0	Yes	2
J66	1	100	0	Yes	2
J140/180	1	20	26	Yes	3
B4	0	—	—	(Yes)	1
B8	0	—	—	Yes	1
B11	0	—	—	No	—
B12	0	—	—	No	—
B39	0	—	—	(No)	—
B71	6	20	0	Yes	1
B145	9	20–5	0	(No)	—
B80	7	30	2	No	—
B120	7	20	2	Yes	1

Note: Parentheses signify borderline categorizations; see text.

high turnout for elections (over 90%), most nominees do not actively seek to be elected. The top three executives are elected directly by all members, and they in turn select nine of the representatives to be on the executive committee (although officially this is done by the representatives themselves). Of the 12 executives at the time of the interviews, one was a section chief, and two were foremen.

This structure serves several functions besides collective bargaining, which is carried out twice a year – the annual round and bonus negotiations in spring, and bonus negotiations again in autumn. It also serves for joint consultation, of which there are two levels. The top joint consultation meeting between senior managers and union executives is convened four times a year to discuss business conditions and plans. There are also monthly meetings at the departmental level between representatives and managers to discuss matters within the departments. Union executives also sit on various joint committees (with management) such as health and safety, wages, welfare and mutual aid (*kyosai*), and they head various union committees and study groups, such as production and productivity, human resources development and shortening of working hours.

Mass meetings, for which there is an 80–90% turnout, are held five times a year at the factory, even though it takes members from the headquarters in Tokyo more than an hour to get there. Those who are absent without permission are given warnings, and repeated offences may lead to fines. Thus there is a coercive element to attendance. The annual general meeting is held in September, where the executive committee gives a report on the last year's activities, and policies for the coming year are presented and debated (sometimes hotly with members who are also members of the communist party).

Union organization at the other large factories was quite similar. At the large, multiplant J140/180, there was an extra tier of senior representatives, extra committees, and a third tier of joint consultation.[2] The main difference at J50 lay in the fact that it had two unions, as well as a separate staff association for the employees at headquarters in Tokyo. The first union had retained about 45% of the membership and was strongest on the shopfloor, whereas the second union was strongest amongst non-manual workers. All the machine shop operators interviewed belonged to the first union; some joining after the split did not even realize there was a choice. The original animosity between the unions seems to have abated, and there were prospects for a merger in 1989. Both unions relate to the company both through collective bargaining and joint consultation, sometimes together.

The union at J66 seemed more lethargic than the others. There was only one union representative for every block of 100 or so employees as well as the chairman, vice chairman and secretary. The present chairman, a chief foreman, has been the chairman for 10 years.

There was no formal industrial relations machinery at the smaller factories, with the exception of J4, where the MD had in fact prodded his employees into establishing a union (actually more like a staff association, despite the use of the word *kumiai*). This was resisted at first by some of the older workers. They were loyal to the deceased founder, and thought the son was trying to distance himself from them, which would make it easier for him to lay them off in a business slump in the future. They preferred a family-like relationship, which the son was anxious to do away with in order to rebuild the company. 'I'm not the founder, and I don't have a technical background. I can't expect the workers to be loyal to me on either of those counts. Family concerns might go well in the first generation, but not in the second.' However, he admitted, 'Just as they couldn't do away with the emperor system and introduce democracy overnight, it takes a while for people's thinking to change.' The situation was different at J2: 'We are suffered to work here,' quipped one worker.

Coming to the British factories, at B71 there were four manual unions (AEU, AEU foundry – treated separately from AEU – Patternmakers – also officially not independent – and GMBATU) and two staff unions

(APEX and TASS). By far the biggest union was the AEU, which organized two-thirds of the shop-floor workers. One steward was elected for every 20 or so members, ostensibly annually, but in practice until the workers became dissatisfied. Said the outgoing senior AEU steward: 'Yeah, names should be put forward and elections held at a full meeting, but you never get a full meeting. I got the job in the first place because no-one else wanted to do it. That's the way things often happen.' He was rather despondant about turnouts for meetings; 'If the miners called a special meeting, 80% would turn up. We would be lucky if we could get 30%. We usually only get about 10 [people] at the [weekly] branch meeting.' These figures were not particularly low for the larger British factories, but stand in marked contrast to attendances in Japanese meetings.

The stewards pick the negotiating committee; three from the AEU, one from the foundry, three from GMBATU and two patternmakers. The senior AEU steward said that the manual unions 'stick together on most things' but the industrial relations manager thought that they were autonomous in most areas except for pay negotiations. An attempt to have manual and staff unions work together lasted for one year and then ended, apparently because information was not being disclosed due to mutual suspicion.

A Works Council, inaugurated in 1983 in the midst of redundancies and low morale to improve communication, consisted of senior managers, middle and line managers, shop stewards *and other employee representatives*, and was chaired by the managing director. It still met every second month, and the industrial relations manager claimed that its effectiveness had increased as managers became more adept at presenting information. Worker suspicions, he claimed, had been largely overcome, but workers interviewed claimed that information was not getting through to them.

At B120, consultation appeared to be an extension of bargaining arrangements. The shop stewards' joint negotiating committee talked to personnel once a week on Wednesday mornings, followed by a meeting with all shop stewards in the personnel department. The personnel department, however, was attempting to set up 'communications meetings' between managers and their subordinates. The stewards had told their members to boycott the meetings, fearing that the company was trying to bypass them. The company responded by informing the stewards of the contents immediately before the meetings. The issue was still not resolved at the time of writing. At the other unionized factories, very basic joint consultation arrangements were being established – twice yearly productivity meetings at B145, for example.

At the other unionized British factories, too, there was quite a lot of cooperation between manual unions, but little between manual and non-manual unions. There *was* a joint staff–trades committee at B145 which,

Table 4.2 *Negotiating/consultation bodies in the larger British factories*

	B71	B145	B80	B120
1 Unions				
no. of manual unions	4	6	4	4
no. of staff unions	2	3	3	3
2 Contacts with outside officials				
often(O)/sometimes(S)/never(N)	S	S	S	S
3 Joint committees (yes/no)				
joint trades	Y	Y	Y	Y
joint staff–trades	N	Y	N	N
combine committee	na	N	N	N
4 Wage negotiations				
manual unions together	Y	Y	Y	Y
staff unions together	Y	Y	N	Y
manual and staff together	N	(N)	N	N
5 Works Councils	Y	N	N	Y
stewards only (S), stewards and non-union (S&N), non-union (N)	S&N	na	na	S

curiously, was the main wage negotiating body for the manual workers, but not for the staff side. The joint staff committee negotiated staff wages separately. The obstacles to creating a unified negotiating body were perhaps most clearly expressed at B80. The white collar unions (APEX, ASTMS and TASS) traditionally acted independently, despite having a joint committee. With the restructuring, the personnel manager felt they would soon find that they had to work together, and that he would be able to negotiate a collective pay deal. (ASTMS and TASS have, in fact, since merged.) That would not overcome the works/staff separation, however. 'Some of the manual unions are still unhappy about sitting down with the staff, who might be their bosses.' The reservations worked both ways.

All this does not mean that the *process* of industrial relations in the Japanese factories was unproblematic compared with the British factories, for as we shall see, there had been major disputes, but there was much less dispute about the *machinery*, both amongst groups of workers and between workers and management. (The potential for manual/non-manual divisions in the Japanese factories was not completely eliminated by single unions, but care was taken to maintain a balance on the executive committee. If the chairman were from a manual department, for example, the vice chairman would be from a staff department. Sometimes there were rotations, sometimes not; either way the balance was recognized in custom.) The same machinery also served more functions. While there were joint committees in the large British factories – safety, pay and grievance – the unions did not have their own extensive sub-committees and study groups, and the issue of whether or not the union organization would serve as the vehicle for joint consultation or whether representatives would be elected directly was not resolved.

While the smaller British factories, including B39, did not recognize any union, there were normally works councils or committees. Some of these were in effect moribund. At B39, for example, there used to be a works committee, but the representatives gave up – because they felt they weren't getting satisfactory answers according to one worker, because they felt the workers were not backing them up according to another. The shop-floor workers spoken to, however, thought they were able to express their concerns directly, and noted that the managing director made a point of going around the shop-floor every morning and talking to them. As at B12, for the managers demands for union recognition would constitute a failure in employee relations.[3]

A union was once organized at B11 by an engineer. It was recognized by the management, and, with the help of an outside AEU official, negotiated a holiday agreement. Eight people on the whole floor were interested, according to the foreman, but they were paying fees and getting no more than the other workers, and they had to go out on national strikes, which did not make them very popular. When the organizer left interest waned, and only one worker was still a union member. There was no joint consultation, either. 'There just aren't enough grievances,' contended the foreman.

Union resources and outside links

The union of J140/180 had 26 full-time executives and 12 full-time clerks, all of whose wages were paid by the union. While there were no full-time officers in the unions of the other factories, they did have 'part-time' office clerks who worked from 9:00 to 4:30 and whose wages were also paid by the union. In the British factories there were full-time officials only at B80 and B120 – one AEU and one TGWU official at each factory (plus a deputy to stand in for them when they were away), whose wages were paid by the company.

J45 published two booklets for its members besides a monthly news-sheet, one concerning negotiations (four times a year) and the other on welfare and social concerns and events (several times a year). It also produced a report of about 150 pages for the annual general meeting which was distributed to all members and covered everything from negotiating letters to committee reports to its survey results and union-sponsored recreation activities. The union at J140/180 was even more prolific, those of

J50 slightly less so, while that of J66 seldom produced its own publications, relying instead on Zenkokukinzoku (to which it was affiliated) newspapers for its members. The unions in the British factories normally only ran off sheets before negotiating time, or when a special issue arose.

The differences are due in part to funding. AEU workers paid 82p per week, which represented roughly 0.6% of their average basic wages, about a third of what the Japanese workers paid (1.6% at J45, 2% at J66), but the latters' fees stayed mainly within the company and went to only one union.[4] Differences in publications, however, were also attributable to the qualitatively different approach to industrial relations on the part of the unions alluded to in chapter 2, which in the Japanese factories meant a greater reliance on written materials, formal 'hierarchical observation' (internal and external surveys) and the power of persuasion rather than the persuasion of power. On the company side, monthly or bi-monthly newspapers and magazines were produced, often carrying a speech by the company president (like a 'State of the Union' address), the current state of sales and order books, QC circle activities and reports, personnel changes, social events and so on. The British companies did produce newspapers – generally bi-monthly – which were partly for customers and partly for workers. They did not contain a 'State of the Union' speech or a detailed breakdown of company performance, although B120 was making elementary steps in that direction with its communication drive.[5]

Most of the funds paid in by the Japanese members stayed within the enterprise union. Where the union was affiliated, some money was passed up; this was one reason why J45's union had withdrawn its affiliation with the federation Domei. Members also thought that campaigning for Domei-sponsored candidates at election time was troublesome. While the second union at J50 was also non-affiliated, the others were, and the executives seemed to have at least as much communication with outside officials as senior stewards in British factories (although this was hard to gauge). Outside officials, however, were never present during negotiations.

Executives of the Japanese unions spent at least two days per year (which they took as annual leave) receiving training at their federation, the prefectural labour department, the Labour University, Japan Productivity Centre, etc. Representatives also received about two days' training per year (although not at J45), sometimes at the company, with outside speakers often invited. The British factories had agreements about day releases for stewards for training, for example by the TUC or AEU. Interestingly, at B120 the company also provided training for stewards. Remarked one manager: 'You get stewards going on a course organized by their group training function on "How to Negotiate With Your Manager" and the managers haven't even been trained for it!' Paradoxically, then, while there was a greater gap between the unions and the employers in the British factories, they were less financially independent in terms of pay for executives, supplies and training costs.

On the whole, though, the organization of industrial relations at the larger Japanese factories and the use of resources by the unions reinforced organization consciousness. It was first and foremost the union of the particular company – one could gather information on company sales, profits and even capital, plant and products from union materials – whereas the organization in the British factories reinforced occupational and status (manual versus non-manual) consciousness more strongly than that of organization, even though the main stage for industrial relations was within the factory.

Mood

That does not mean that the process of industrial relations was always smooth in the Japanese factories and confrontational in the British factories, or that there had been no challenges for control of the union organization in the former. The history of industrial relations in the larger Japanese factories has been one of gradual 'stabilization', starting before but particularly following the first oil shock. 'Stabilization' may also characterize the more recent history in the larger British factories following large-scale redundancies, although there are important differences. Let us start with the Japanese factories again.

As mentioned in the last chapter, the union at J45 was formed in 1954 over the lowering of piece rates. This happened again in 1959, which resulted in an overtime ban for about a year, and reduced communication between the union and the management. An uneasy peace was maintained until 1973, when the new pay system was initiated. That did not result in a sudden improvement in industrial relations, however. In fact, in that year there was a dramatic swing in the opposite direction.

Members of the communist party were voted on to the union executive committee which, during the subsequent oil shock, demanded a ¥30,000 average wage increase. The company refused, there were strikes and a court injunction, and the company finally settled for half the original figure. Negotiations the following year were even more acrimonious. Flags and posters were put up, there were pickets, workers lying down in front of trucks and fighting and shouting. Facing heavy penalties for late deliveries, the company virtually capitulated, but managed to get the top union leadership and several other representatives suspended. This further soured relations and there were sporadic strikes until 1977 when the workers, apparently tired of strikes, voted out the communist leadership. There were still dissident workers handing out fliers at the gate of the local railway station in 1987, an irritant at most for the personnel management, who said that most members knew what the fliers said without reading them, anyway.[6]

Present industrial relations are very cooperative. The union even has its own study group on productivity, as noted. According to a recent union survey, 31% of members think their views are not reflected in union activities as against 56% who think they are. Some of those interviewed thought the union was too close to the management, but felt that was preferable to the situation in the more turbulent days.

Industrial relations at J50 and J66 had also followed chequered paths. At J50 the union pushed for a ¥20,000 rise for its members in 1975. The company refused to pay and in the ensuing strike the union split. Sporadic strikes by the first union continued, but the split had weakened it. It did, however, successfully take the company to court in 1982–3 over discriminatory pay rises against its members. Since then, relations have stabilized and 'efforts have been made by all sides to improve communications' (personnel staff). The unions are set to merge again.

When the former president was brought in by a bank to rescue J66 in 1950, one of his first acts was to lay off 30% of the workforce, for which he found himself surrounded by 2,000 irate unionists, mobilized by the federation Sohyo, to which the union was affiliated. This president evidently did not like collectivist principles, which might account for the underdeveloped state of the union, but he accepted that he had to live with it. He took pride in the fact that no-one had been fired since then, except a union secretary in the early 1970s, who threw an ashtray and injured the general affairs manager during a heated bargaining session. Again, this was in the post oil-shock period, and relations had subsequently become more cooperative.

There was also a strike at J140/180 in 1950 when almost 40% of the workforce were laid off, but none since, a fact of which managers are very proud.[7] The union apparently took note of disputes in other firms at the time of the first oil shock, and made only moderate demands.

This illustrates the diverse and sometimes troubled industrial relations histories in the larger Japanese factories. There had been, however, a process of gradual stabilization. Major changes in employment, payment and industrial relations had been in place for a number of years, and these were being fine-tuned. More recent changes had been at the interface of industrial relations, personnel and production, particularly QC circles and related activities, which will be discussed along with other kinds of innovation in chapter 5. Some of this 'stabilization' of industrial relations, however, must also be attributed to the constriction of manoeuvring room in bargaining which the unions have experienced since the mid-1970s with the onset of slow growth.[8] This constriction is a major factor in leading unions to seek a bigger role for supercorporate union organizations, particularly the new national centre Rengo (Japanese Private Sector Trade Union Confederation) according to some observers.

In the smaller Japanese factories where there was no formal industrial relations machinery (except at J4) there were differences in management style which affected the industrial relations 'mood'. The managing director of J9 thought the company was still small enough to obviate the need for formal industrial relations machinery, but he did note that he was trying to parcel out responsibility and prepare the company organizationally for rapid expansion in their new premises. Nonetheless, he did sometimes sense dissatisfaction about his company being family-run. The managing director of J4 was seeking to move from personal discretionary control to a system of authority founded in a more egalitarian corporate consciousness. He had purged all other relatives from active positions (although his mother was still the largest shareholder): 'I want to show the workers that we [the top managers] are just another function like them, side by side and not above them.' Morale, he said, is created by raising people's consciousness, not by indoctrinating them. He cuts items of interest out of newspapers, pastes them on paper and circulates them in the office and in the canteen. In order to get his workers to accept evaluations positively, he first had core workers evaluate the directors on how clear their directions were, how hard they were trying to understand workers, how responsible they were towards their work etc., and then made the overall results known.

J2 and J1 are still family affairs, managed by personal control. Their common concern was to keep wages high enough to secure compliance, but not so as to raise the proportion of labour costs too high. There were few attempts to foster value congruence – although there was a small core of long-serving, loyal workers at both – and the eldest son at J2 admitted that they awaited for signs of commitment on the part of workers before they went out of their way for them.

The British factories, too, had seen a 'stabilization' of industrial relations of sorts. At B145, which was in an area noted for worker militancy, the issue of union control over the work process came to a head in 1978. 'To lift a job on and off a machine needed three people. Guys were spending half a day up in the cranes doing nothing. You had to get permission from the unions to do overtime. Customers were hanging on the telephone. Will I get my order or not? Nine times out of ten we couldn't get permission. Things just couldn't go on that way' (department manager). The unions also insisted that supervisors should not work on the machines: 'We couldn't even operate machines for instructional purposes' (superintendent).

In that year there was a 4-week strike over wages, which the workers lost.

The management used the occasion to force through an agreement on flexibility and the productivity scheme it had been proposing. Amongst other things, the agreement (with the Joint Trades Committee) stipulated: 'The acceptance in principle and practice of new technology and other capital equipment. Any such equipment will be operated and maintained to its maximum potential immediately it becomes available'.

Apparently the managers did not have the resolve to push the agreement through to fulfilment, and old practices crept in again. It was not until another strike and management victory in 1984 that they found this resolve 'because we had been through the redundancy crisis of 1980, and we were determined to make it work'. (In 1980, 600 workers were laid off, a very painful business which nevertheless was carried out without a day's stoppage, of which the industrial relations manager was very proud.)[9]

The convenor was philosophical: 'Yes, there used to be restrictions on subcontracting. We used it as a bargaining tool. Management got browned off. That went in '78, but there hasn't been an increase in work subcontracted out. As for time and motion studies – the lads know the jobs best and can fix the rates. I can count on my hand the number of times they've been in since then. And flexibility – they don't want skilled turners running around with paint brushes for too long... They won their case; they have to look like they're in control.' He thought that on the whole industrial relations were 'pretty good'.

One department manager thought they were the best ever. He described how some operators had recently been asked to go and do jobs in a store room and were delighted at the change when they would have been absolutely against it a few years back. Why the change? 'Well, for one thing, a lot of their fears have been overcome. They used to be worried about pay – maybe they would be stuck there and their pay would be docked, and they were worried about failing something new'.[10]

There had been very little major industrial action in any of the larger factories in the past three or four years. Undoubtedly this was related to past redundancies and the possibility of more with pressures of competition. But are the recent changes no more permanent than a phase in a see-saw game, which is how Fox (e.g. 1985a, 429) has viewed them? Anecdotal evidence does not give a definitive answer, but there were indications of a more permanent change.

Many of those interviewed thought that the younger workers belonged to a different generation. This took the form of scepticism at the traditional attitudes of the older workers, scepticism about their unions, and sometimes apathy. 'The union? They're the older people.' 'Me standing? No, I'm not union material. Most of the others my age aren't bothered, either' (young operators, B71). 'The union's getting eroded away. A lot of the senior stewards have gone, and it's a case of David against Goliath now. Goliath wins every time' (senior operator, B71). 'The union? It used to be too militant. I voted against the overtime ban. I guess more people would have stuck with the union before. People think for themselves more now' (operator, B145). 'They look at the older ones and think is that what's in store for me? They're anxious to improve themselves. They come and talk to the company about further educational opportunities' (industrial relations manager, B145).

As we have seen, personnel policies were evolving, too, if slowly, towards 'human resource management'. Sometimes the talk of change was more rhetorical than descriptive, but most agreed that changes in this direction were on the way, and that again, quite a few younger workers expected ongoing training and mobility. The unions at B120, which were still comparatively strong, were also increasingly supportive of ongoing training since it enhanced the job security of their members.[11] Harmonization, attempts to improve communications and increase identification with the company have also been mentioned, and although many of these represented piecemeal responses to problems accruing from the past, their combined effect may be to change the nature of the 'game', though not as fundamentally as in Japan.

It was in the non-unionized factories like B39, B12 and B11 that managers talked most freely about recruiting motivated workers who wanted to get on, using internal promotion, evaluating and rewarding workers' efforts. While some of this was also more rhetorical than actual, there are fewer structural impediments to bringing about such changes, although fewer resources such as promotion opportunities as well.

Industrial relations in the larger British factories could hardly be described as 'cooperative', as distinct from not overtly confrontational. The different groups still had their own interests to protect. While unions had come to be perceived by managers in the larger Japanese factories (with the possible exception of J66) and even at J4 as valuable instruments for organizing and integrating workers, they were perceived more as obstacles to change or at best a fact of life in the British factories.[12] A large number of Japanese company managers and executives had once been union executives or representatives. At J45 the company president was a former union executive, as was the general affairs manager. The factory manager was a former union secretary, and the personnel manager was once the chairman. The technical director at J66 had been the deputy chairman, and two of the other directors had been representatives. This was not accidental. Explained a personnel manager at J140/180, where the company president was also an ex-union executive: 'Union officials who show ability in balancing and presenting the interests of the members have good management potential.' The personnel manager of B71 was a highly-rated ex-steward, but this was exceptional in the British factories, particularly as far as manual unions were concerned. Stewards might be candidates for foreman posts, but the requirements of being a foreman and of being a senior manager were not the same.

Just how much these factors translate into differences on the OER–MER scale, we will soon see, but before that there is one more issue to be discussed in the context of industrial relations.

New technology

The introduction of new technology in all the factories was a management prerogative, and was communicated to a greater or lesser extent to the workers through informal and sometimes formal channels.

Formal channels in the large Japanese factories meant reports of capital expenditure and not joint consultation channels. Workers' opinions were solicited informally, however. At J45, for example, the union secretary was the programming expert, and his ideas were solicited in this capacity before CNC machines were bought. A senior operator, who was also a union executive, said they heard about the FMS system about a year before it was installed when they were asked by their superiors what equipment they thought was necessary. He said that the union was not worried about the introduction of advanced technology because of a 'trust relationship' with the company, and referred several times to the policies of the company president stressing the need for creativity and advanced skills with advanced automation. The union secretary put it thus: 'We are still in favour of bringing in machines like CNC. The number of people wanting to work on them is high. The policy of the [company] president, too, is that workers shouldn't become like robots, that they should have human [*ningenrashii*] work, so we are in basic agreement there.' He was in the process of organizing a study group on microelectronics innovations to work towards the goal of a 'workplace worth working in' (*hatarakigai no aru shokuba*).

At J140/180 the planning executive of the union argued that the union could not oppose new technology, and according to a personnel manager, the union has pushed the company to automate more, to maintain the present workforce and raise productivity. They also hoped that new technology would reduce the present workload. 'The union here isn't concerned about unemployment, but it is concerned with a fair distribution of the fruits of increased productivity,' he added. Of recent union reports, only one made a brief mention of new technology: 'With the advance of technological innovation and the introduction of microelectronics machines, we have proceeded by holding joint talks placing emphasis on safety, as well as changes in job contents, the workplace environment, and so on.'

In the smaller Japanese factories there was even less management–labour discussion on new technology. Purchases were decided by the managing director or the family, although the foreman was consulted at J4.

In the large British factories there was some dispute about information provision. The manual unions at B145 had no established policy regarding the introduction of new technology, according to the convenor. They were basically reactive, he said. When they were still powerful a machine would have been blacked if it were brought in with no prior consultation (and indeed, that had happened). The clause in the productivity agreement regarding new technology was merely to ensure that its introduction would not be hindered. 'The management like to give the impression they're managing', said the convenor, with the implication that they gave as little prior warning as possible (this would indicate greater strength where strength was an important factor in the process of industrial relations). The operators knew about the giant new machining centre that had just come in 3–4 months beforehand from the twice-yearly departmental productivity meeting.

By contrast, the company did have an agreement with TASS about the introduction of new CAD equipment dating back to 1979, which granted an extra £4 per week for those working on CAD (in contrast to the stance of the manual unions). This agreement was expanded in 1981 and guaranteed that no-one would lose their job involuntarily as a result of the introduction of new technology; those displaced would be redeployed, salary levels would be maintained, training provided and re-evaluations carried out according to agreed procedures. Moreover, 'the Company will involve employees and their representatives before changes are implemented. For example: (i) to agree provisions for a training programme; (ii) to discuss the identity of jobs affected, new or changed job descriptions, and the nature and timing of change.'

At B71 there was also an agreement with TASS but none for the manual unions. According to the industrial relations manager: 'The company has always given information of the capital equipment it intends to buy, even before the days of the Works Council. We have never had any problems in obtaining consent. We have not had a single problem with CNC, although there are some companies I could tell you of where machines have stood idle for six months because of disputes.' Some of the senior operators, he pointed out, including the outgoing senior steward, had gone to manufacturers with the production engineers for the final selection. Again, some of the operators had a different view. 'We had a heck of a fight. The management might have had skilled setters only if they could have gotten away with it.' Information from the Works Council was allegedly not getting down to the shop-floor. Whichever view was correct, no new technology agreement with manual unions had been discussed or signed, in contrast to one which had just been concluded with TASS regarding CAD/CAM, portions of which had been discussed since 1975.

At B120, where problems had arisen over the introduction of new technology in the past, some effort was being made to include it in the communications drive. By and large the manual unions had taken a cooperative stance towards microelectronics equipment, and like those of B80, were becoming increasingly keen to have their members receive as much training for it as possible (see chapter 6). None of the workers interviewed in the larger factories attributed redundancies to new technology. There was little indication that new technology was a matter for consultation in any of the smaller factories, except for informal discussions with foremen and senior operators.

4.2 OER, MER and the 18 factories

The final task in this discussion of employment relations is to evaluate the factories on an OER–MER scale. This will summarize the findings, as well as provide a measure with which to relate employment relations to aspects of CNC use. Rating the 18 factories on a continuum between the poles of OER and MER is difficult for several reasons. First, if one relies on quantifiable data only, contextual factors necessary to interpret the data will be missed. Recruitment, for instance, cannot be considered independently of the economic 'climate'. Secondly, how does one deal with divergences between stated goals and measurable indices, as we saw to be the case for mobility in both countries? A factory with a policy of rotating workers as part of career development may actually have a lower count of job mobility than one which allegedly only shifts workers in response to immediate production needs. Thirdly, what weighting should be given to the various factors? Should industrial relations factors be given more weight, the same, or less than those of payment, for example?

Both statistical and interpretive criteria are used in the evaluation. The position is taken that OER involves the conscious pursuit of goals and objectives, for example those concerning worker development, which are communicated to workers as company expectations. Moreover, the closer shop-floor realities approach these, the more organization-oriented the factory in these respects. The scoring gives equal weight to employment, payment and industrial relations factors, although one might argue that an employment factor like recruitment is more subject to environmental influences, or that payment as the encapsulation of company expectations towards the worker should be given greater weight. All three are crucial and mutually reinforcing, and in fact some of the criteria can be considered under more than one category.

Five criteria have been selected in each category, and each factory is ranked from 1 to 5, with 5 representing the organization pole and 1 representing the market pole. The higher the total points, the nearer the factory overall to the OER pole. Where different ratings apply to different groups of manual workers in the same factory (e.g. apprentices and non-apprentices in number 1), or where there is more than one component to the evaluation of a criterion (e.g. recruitment and pay systems in number 2) and there is a discrepancy in these, an average of the scores is taken. Again, it should be remembered that the principal reference group is manual workers.

EMPLOYMENT

1 **Recruitment**: The vigorousness of the recruitment process of manual workers.
2 **Mid-term entry**: The degree to which mid-term entry is discriminated against in both recruitment and pay systems.
3 **Induction**: The amount of time taken for initial orientation, indicative of the company's desire to orient and socialize recruits into the norms and goals of the organization.
4 **Fixed employment**: The proportion of workers under 'fixed employment' (beginning at age 25).
5 **Career**: Interpretation of the degree of penetration of personnel policies for viewing workers (a) as resources to be developed, and (b) as being employed for a career in the company.

PAYMENT

6 **Orientation**: Fixed intra-corporate rules relating to pay rises and relativities (with reference to market rates only, for example, as starting rates, average wages and increases) versus frequent reference to market rates for specific skills/jobs.
7 **Flexibility**: The ease with which workers can be moved from job to job, or job contents changed, without considering pay implications.
8 **Evaluation**: The degree to which the contribution of the individual towards organizational goals is recognized, and the degree to which it is recognized by formalized criteria.

9 **Company performance**: The inclusion of company performance as a criterion in wage negotiations, and its reflection in bonus payments.
10 **Harmonization**: The degree to which all regular employees are recognized as full members of the organization in payment and conditions.

INDUSTRIAL RELATIONS

11 **Machinery**: The presence of industrial relations machinery, particularly joint consultation, and the extent of its use.
12 **Level**: The congruence of the focus of industrial relations with that of personnel, business and corporate planning functions.
13 **Inclusiveness**: The inclusiveness of worker organizations, and the degree to which they correspond to the boundaries of the organization.
14 **Communication**: The quantity and quality of communication channels, both verbal and written, which may foster a sense of common goals amongst workers and managers.
15 **Common destiny**: An interpretation of the extent to which workers and managers see their futures as intertwined with and within the organization, based on interviews of both groups.

The scoring system favours larger factories in certain respects, which reflects the reality that larger factories have more resources to sustain OER. Besides such criteria as numbers 5 and 6 (career and orientation), this reasoning can be seen in the industrial relations criteria, such as numbers 11 and 14 (machinery and communication). Size alone, however, is not a sufficient condition for OER. In numbers 12 and 13 (level and inclusiveness) organization-type industrial relations will score the highest, formalized trade-based industrial relations the lowest, and those factories without formalized industrial relations will score in the middle; the reasoning being that while formalized organization-type industrial relations support OER, formalized trade-based industrial relations may positively inhibit it, and a lack of formalized industrial relations, while not promoting OER, will not inhibit its formation in other areas.

It was also reasoned that increasing factory size requires increasing communication and commitment to communication initiatives in order to maintain a constant level, and this was taken into account in scoring so that a large factory needed more formalized channels to get the same score as a smaller one with less.

The fact that such very different factories as J2 and B80 have the same scores, despite being incomparable in many ways, demonstrates the limited descriptive or explanatory utility of broad categorizations, but a number of useful observations can be made. First, OER becomes more and more pronounced with increasing size in the Japanese factories. The smallest ones are quite near the MER pole even relative to the British factories. J4 and J9

Table 4.3 *OER–MER scores*

Factory	J1	J2	J4	J9	J45	J50	J66	J140	J180
Criterion									
Recruitment	1	1	2	3	4	4	4	4	5
Mid-term entry	1	2	4	4	4	4	4	5	5
Induction	1	1	2	3	4	4	5	4	4
Fixed employment	2	1	3	4	4	4	4	5	5
Career	1	1	3	4	4	4	4	5	5
Orientation	1	1	3	3	5	5	5	5	5
Flexibility	3	5	5	5	5	5	5	5	5
Evaluation	3	3	5	4	5	4	5	5	5
Company performance	3	3	4	3	5	5	5	5	5
Harmonization	4	2	4	4	5	5	5	5	5
Machinery	1	1	3	1	5	5	4	5	5
Level	3	3	5	3	5	2	5	5	5
Inclusiveness	3	3	5	3	5	2	5	5	5
Communication	3	3	3	3	5	4	4	4	4
Common destiny	2	1	3	4	4	4	4	5	5
Total	32	31	54	51	69	61	68	72	73

Factory	B4	B8	B11	B12	B39	B71	B145	B80	B120
Criterion									
Recruitment	1	2	2	2	3	3	3	3	4
Mid-term entry	1	1	1	1	2	2	2	2	2
Induction	1	1	1	1	1	2	2	2	2
Fixed employment	2	2	3	4	3	3	3	2	3
Career	1	2	2	2	3	3	2	3	3
Orientation	1	2	2	2	2	2	2	2	2
Flexibility	3	3	3	4	3	2	2	2	2
Evaluation	2	2	3	3	4	1	1	1	1
Company performance	1	1	1	1	1	3	3	1	3
Harmonization	2	3	4	3	3	3	3	2	2
Machinery	1	2	1	1	2	4	3	3	3
Level	3	3	3	3	3	2	2	2	2
Inclusiveness	3	2	3	3	3	1	1	1	1
Communication	3	3	3	3	3	4	2	2	3
Common destiny	2	3	3	3	4	3	2	3	3
Total	27	32	35	36	40	38	33	31	36

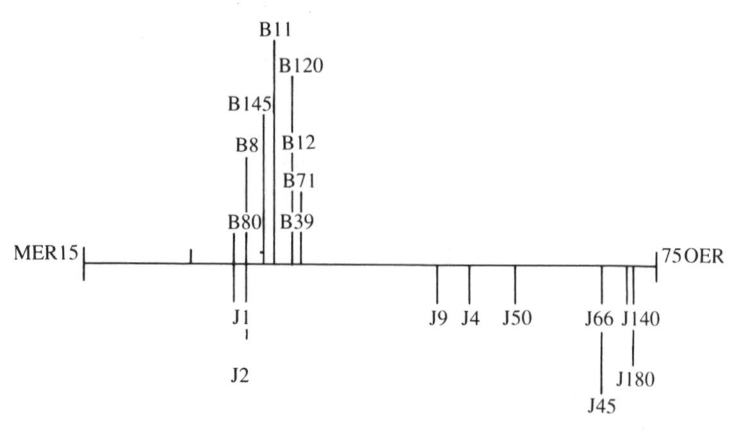

Figure 4.1 OER–MER continuum

are perhaps nearer the OER pole than many other Japanese factories of the same size might be because of the vigour and methods with which the managing directors were pursuing rejuvenation and growth strategies. While the same trend towards OER is observable in the smaller British factories, it is halted and reversed to some degree in the larger factories, which may be attributed to increasing formalization of MER, although the trend is not nearly as pronounced as it would have been had the calculations been made 10 to 15 years ago.

Perhaps the greatest differences between the British factories and all but the smallest of the Japanese factories was the degree to which *all* regular employees were viewed as (a) potential resources to be developed, and (b) as having a career within the company. Most telling in this respect was the common expression (by managers, regardless of factory size) that 'the cream gets to the top'. None were committed to the belief that all workers were potential cream (as a result of the selection process), or that an important aspect of the employment relation was bringing out the 'cream' in all workers, which would closely represent the stated beliefs of Japanese managers of all except the two smallest factories (although privately, they, too, probably took a lower view of the potential of some of their workers, but were careful never to admit it).

The scores of the factories can be shown on a continuum, with a score of 15 representing the MER pole and 75 representing the OER pole (see figure 4.1). The rankings of the factories from the scores will be referred to later, when they will be correlated with rankings on aspects of CNC use – training and task ranges. The main weight of evidence, however, rests with the description given in chapters 3 and 4, and the same will apply when discussing the hypotheses in subsequent chapters.

5

Innovation

Two aspects of the move to CNC – pay and prior consultation – have been looked at briefly. In this chapter I shall go on to look in more detail at the introduction of CNC into the 18 factories. First I shall look at when CNC was introduced, and to what extent. While the large British factories were the first to purchase CNC, or NC, the Japanese factories had caught up and in some cases surged ahead in the number of CNC machines purchased. More noticeable than numbers of CNC machines, or even the linkage of these into more advanced forms of automation, however, were the different views of 'best use'.

The following section looks at another type of innovation; small group activities – particularly QC circles – and JIT (just-in-time) initiatives. Warner (1984, 203) warns that 'if new technology is introduced into firms which do not adopt new structures, then the costs of implementation may be high and the potential benefits diminished'. The implication is that many British factories have been lacking in this respect. In Japan, on the other hand, the introduction of new technology may be viewed more 'wholistically': 'Many Japanese companies view the adoption of J.I.T. [just in time] as the prelude to full factory automation to further reduce costs. . . Factories must be running efficiently before they can be automated' (Abegglen and Stalk, 1985, 116–17).

Such differences – if they exist in our factories – will have implications not only for productivity and efficiency, which are not being discussed directly here, but also for training and the division of tasks and CNC use in general. The final section looks at the extent of linkage between these types of process innovation and whether these in turn are linked to employment relations.

5.1 Introducing CNC

Numerical control (NC) is a technique by which instructions are fed into a machine in the form of a code which consists of numbers, letters of the

Table 5.1 *Introduction of CNC*

Factory	First NC/CNC purchased	No. of NC/CNC	Percentage	Factory	First NC/CNC purchased	No. of NC/CNC	Percentage
J1	1982	2	18	B4	1980	3	6
J2*	1975	6	13	B8	1984	4	23
J4	1979	7	45	B11	1982	3	13
J9	1978	28	41	B12*	1975	17	45
J45	1973	15	17	B39	1970	6	10
J50	1971	3	3	B71	1972	18	15
J66	1973	10	13	B145	1958	39	11
J140*	1976	181	12	B80*	1967	16	5
J180	1970	100	25	B120	1961	30	15

Note: Percentage means the proportion of NC/CNC machine tools to total machine tools in the machine shop area(s) excluding conventional machine tools 'never used.'
*Indicates large batch factories

alphabet, punctuation marks and other symbols. The machine responds to this information in a precise and ordered manner to carry out various machining operations.[1] J.C. Parsons of the United States is generally considered the father of NC, starting with his punched card Digiton in 1947, with the first continuous path NC milling machine being demonstrated at MIT in 1952 (the process and politics of which are described in great detail by Noble, 1984). Similar systems were being worked on in Britain by such companies as EMI, Alfred Herbert and Ferranti, which demonstrated the Ferranti numerical measuring system in 1951. Early NC control systems were 'hard-wired', which meant that any options had to be specified in advance, upgrading was difficult, and many soon became obsolete. The machines were bulky, the controllers expensive, and the tapes took considerable time to produce and could be damaged by the heat and dirt of the shop-floor. Early purchasers of NC, therefore, were confined to larger firms.

The development of CNC in particular opened the way for wider use, including smaller firms. 'Soft-wired' CNC retains the fundamental concepts of NC, but the control unit houses a stored-programme computer, and many of the control unit functions can be placed on reprogrammable software, which greatly simplifies the hardware, reduces costs and improves reliability. Further developments in control functions, such as MDI (manual data input) means that instead of having to utilize tapes, information can be punched directly into the controller, changed there, stored, etc., and with background programming, this can be done while the machine is still performing cutting operations. Related developments in hardware, such as the development of machining centres, has meant that operations which could formerly only be done on separate machines can now be performed on one machine, with tools automatically changed according to the programme. These developments have increased the utility of CNC for smaller firms, and they have become more affordable as well. According to a JMTBA survey, 58.9% of domestic CNC sales (presumably by value) were to small and medium-sized enterprises in 1987, with 16.7% of the total going to firms with less than 30 employees.[2]

Knowledge of NC entered Japan in 1952, and the first experimental NC lathe was produced at Tokyo Kogyo University in 1956, with Fujitsu developing a tape controlled turret punch press in 1957. The first commercial NC machine was produced the following year. The CNC breakthrough in Japan came in 1969–70, again by Fujitsu. Although it ended the war with a huge technological gap, Japan produced its first NC machine just five years after MIT, and was only one year behind the United States with the production of machining centres and CNC.

Initial purchase

Of our 18 factories, the earliest purchasers of NC were, not surprisingly, the larger ones (see table 5.1). It was not until the mid to late 1970s, when CNC became commercially viable, that the smaller factories made their first purchases. B145 made the first purchase – in 1958 – three years ahead of B120 which was six years ahead of B80. B145 was a full twelve years ahead of the first Japanese factory in purchasing NC. In contrast, most of the smaller British factories came to CNC later than their Japanese counterparts.

The date of initial purchase is important because it influenced both attitudes towards NC/CNC and its use. The first NC machine purchased by B145 in 1958 was a drilling machine, followed by some gun drills in the early 1960s. NC milling machines were first purchased in the mid 1960s, and NC lathes in the late 1960s. Apparently salesmen emphasized how NC would reduce reliance on traditional craft skills. This notion (also common in the US) – plus the state of NC technology – no doubt influenced or reinforced Braverman's formulation of his deskilling thesis in his influential book (1974), and still exists to a certain degree. B145 was at first convinced that little training would be needed for NC operating since it would be so simple, and that NC could be operated by semi-skilled workers. Moreover, the first machines – drills and certain milling machines – were conventionally classed as semi-skilled machines. In factories like B145, however, engineering workers were most strongly represented by the then AUEW. There was the potential for conflict, but as it turned out, this rarely materialized.

Managers decided that 'trained monkeys' were not going to get the most out of their expensive purchases. Subsequent trends were to use skilled operators.

B145 first had its tapes made outside the factory, which was obviously unsatisfactory when corrections had to be made. The programming section grew out of the jig and tool design section. With recent MDI and background programming developments, much of the programming has been delegated (or 'returned') to the shop-floor operators at B145. The history of introduction in the other large factories was quite similar, although they never had their tapes made outside, and they had kept programming to a larger degree in production engineering departments, citing downtime of machines and shop-floor noise and distractions as reasons against operator programming.

The smaller British factories made their first purchases later, often visiting larger factories in the vicinity before deciding not only what type of machine(s) they would purchase, but what kind of operators they would use and how they would organize tasks. They tended to use skilled operators from the beginning, and many also opted not to have programming done at the machine, keeping it instead nearby in the foreman's office.

The larger Japanese factories made their first purchases later than the larger British factories. The first of the nine was J180, which bought an NC milling machine in 1970. There was no production engineering department at the factory then, but in common with a number of other larger Japanese factories, the machine was first put in an area isolated from most of the machine tools, set up by the equivalent of production engineers, and handed over to the machining department when the technology was no longer new. After 1970, J180 purchased one or two NC machines per year. Programming continued to be done in an office, but by 'blue collar' workers under the same supervision as the operators. In Japan, NC was also sold 'to make up for a lack of skilled labour', but J180 said that young but skilful operators had been assigned at first.

There was not a marked difference in the initial purchase dates between the larger and the smaller Japanese factories. J50 bought its first NC machine in 1971, but it was not until 1985 that a second was purchased, well after even the smallest factory had introduced CNC. J2 was the first smaller factory to purchase NC – a lathe – in a gamble after the first oil shock. This was one year before J140, also producing large batches, which surprisingly did not purchase its first NC machine until 1976, although it soon acquired a large number with heavy investment in the late 1970s and early 1980s.

While percentage figures should be treated with caution – what machine tools are included in the overall figures is problematic, for instance – the figures in table 5.1 do suggest that there is not a striking quantum difference between the proportion of CNC to total machine tools in the British and Japanese factories. The larger Japanese factories, however, have been more active in purchasing CNC in the last 10 years, with J140 surging ahead of its British counterpart B80 and now possessing ten times as many CNC machines, and J180 with three times as many as B120. J66 in 1989 was embarking on a major capital investment programme leading up to its 70th birthday, and by 1990 or 1991 will probably treble or quadruple its number of CNC machines. Even J50, long convinced that CNC was not for it with its very small batches, will now purchase at least one CNC machine per year in the foreseeable future.

The smaller Japanese factories also seem to be purchasing more rapidly than their British counterparts, which, with the exception of B12, seemed to be consolidating their experience with their three or four CNC machines and to be unable to justify further purchases with current order levels. J9's business strategy involved large-scale investment in 'high tech' production machines (as did B12's), and J4 was set to expand its CNC plant with the opening of its new factory. Nonetheless, one cannot say that all Japanese factories are bristling with the latest technology while British factories all languish through a lack of capital investment. Several Japanese factories visited had too few CNC machines to be included in this study.

Decision making

A detailed account of the decision-making process involved in NC/CNC purchase is beyond the scope of this book, but some observations are in order. First, many of the Japanese managers interviewed said that 'that was the way things were going' or 'it was the age of NC' when asked why they first decided to purchase NC or CNC. They then decided on the basis of that initial purchase whether or not NC/CNC was suited for their operations. This applied also to other factories not included in the final selection, some of which had had initial failures. Even at as early a purchaser as J50 (1971), a production engineer reflected, 'Well, we thought it was about time (*soro soro*) we got a NC machine.' After buying it, they decided that NC wasn't suited for their operations and did not purchase another one for fifteen years. The improvements manager at J45 produced detailed notes from the study group they had set up to study NC before their first purchase. The notes showed that they had all done considerable prior research, but this was not typical of all the factories. The smaller British factories also seemed to be very anxious about not getting left behind. 'Everyone had CNC except us,' said the CNC 'champion' at B11. Some of the managers at B8 also felt that they were getting left behind after a lack of investment by the previous owners and the opposition of the old MD to anything that remotely suggested numerical control.

Secondly, in the Japanese factories the initial suggestion often came from top management, and the production engineers were left to make a study of which NC or CNC machine would be best for them to purchase. None of the production engineering managers mentioned of their own accord having to write detailed justifications citing pay-back periods, although when specifically asked they said they did do them, and aimed at a pay-back period of four or five years (J45) or even longer (the depreciable 12-year limit at J66!).

In the larger British factories, on the other hand, and even in some of the smaller ones, pay-back periods (usually around two years) were often mentioned spontaneously. Part of the job of the production engineering manager was to come up with these figures, even though everyone knew that it involved 'cooking the books'. 'We never hesitate to buy new technology if we can show a pay-back of two years. That might involve arranging the figures a bit.' Justifications included labour cost savings: 'Saving labour is a major factor, but not the only one. If I can show we're going to save time on jigs and fixtures and programming, that speaks for itself. And replacing machines on a regular basis if necessary is important' (B145).

With the limited information obtained on decision making, however, it might be said that while the *form* of the decision-making process was different from the Japanese factories, whether the actual *substance* was substantially different is an open question. It is also difficult to say whether workers were involved more or less in the decision-making process in the Japanese factories. They were not involved, it would seem, through formal consultation channels, and in many cases they were not informally involved, either. Senior operators were involved on occasions in the larger Japanese factories but in the small ones it seemed that the decisions were made more autocratically.[3] On the other hand in the larger British factories, decisions were made within the production engineering departments, and operators and line supervisors complained that had they been consulted some bad decisions would have been avoided, while foremen and sometimes key operators were consulted in the smaller factories.

Automation

NC/CNC machines may be used in more advanced forms of automation, such as (1) DNC (direct numerical control), where programmes can be downloaded from a computer (in an office) directly to CNC controllers; (2) CAD/CAM (computer aided design/computer aided manufacturing), where design dimensions can be translated directly into a machining code and fed into a CNC controller; and (3) FMS (flexible manufacturing system), where more than one CNC machine is fed automatically by a robot or a cart, which picks up the parts or replaces them according to a predetermined programme.

In smaller factories in both countries, CNC machine tools were introduced on a stand-alone basis. There had been no attempt to link them with other machines, and no robots were used. Machines at some factories, however, were equipped with tables whereby several work pieces could be fixed on pallets and machined in sequence automatically. These included J2 and B12, which were involved in large-batch work, and also one machining centre at J4, and several machines at J9.

By and large, most of the machines at the larger factories were also used on a stand-alone basis, although most had at least one or two machines with pallet tables. Plans for a mini-FMS were well advanced at B71, with a track already laid in front of one machining centre, but there was no pallet changer cart to run along the rails, and no second machining centre for the rails to link up to. The machine shop manager had led a group for two years studying FMS and JIT – the two were closely linked, he argued – but the project was shelved while the company's new owners looked at product rationalization.

At B120 about a quarter of the CNC machines could be programmed directly from the office by DNC (direct numerical control), and for some large parts there was a direct CAD/CAM link between draughting and machining. Although B120 invested heavily in CNC in 1984, however, and had reorganized the machine shops to carry out 'cell' or 'modular' production of families of parts in 1986, the machines were still run on a stand-alone basis.

Of the larger Japanese factories, J50 with its three CNC machines was not even contemplating more advanced forms of automation. J45 began its drive towards fuller automation in 1977 with a machining centre with a 10-pallet table, and ten years later had three CNC machines which were fed automatically by robots (FMC – flexible manufacturing cells) and a mini FMS much like the one B71 wished to build. It comprised two machining centres fed by a programmed cart on rails which took the parts to be machined out of a rack (three levels high and ten metres long) and replaced them again after they were machined. The CNC machines at J66 were used on a stand-alone basis, but a FMS with six CNC machines was due to be installed in 1989 (at a cost of ¥2 billion). Experiments with DNC were also being carried out.

The state of automation at J180 was similar to B120, with limited use of DNC and CAD/CAM. At B80 there was little evidence of more advanced automation beyond pallet tables of two or three pallets, while at its counterpart J140 most of the machines were run individually, with no DNC, CAD/CAM or FMS. About 40 of the CNC machines, however, were connected by transfer machines.

Unmanned factory?

The most striking difference with regard CNC introduction was not so much in the numbers of machines, purchase decision-making process, or even degree of automation, but in the views of managers towards the machines, expressed most clearly with respect to unmanned operating. The differences were drawn along national lines rather than factory size.

'What would you think,' the production engineering manager of B120 asked one trusted operator for my benefit, 'about leaving this £250,000 machine to run unmanned?' 'I wouldn't think much of it at all. I'd get more ulcers than I already have.' The production engineering manager nodded in agreement. At B145, the production engineering manager shook his head; 'No, unmanned operating is not feasible. You wouldn't be able to get the most out of the machines because they would have to be run at slower speeds, and diagnostic options are too expensive. You still get the most out of a machine by having a skilled operator there.' The production engineering manager of B71 agreed, although he did think unmanned operating was the direction things were moving in. The views of line managers mirrored those of the production engineering managers. CNC machines were considered just too expensive to leave them to run by themselves, except for spells during lunch time or between shifts when a piece was being finished off. If the operator did not feel happy about leaving the machine running, he would not, and was not expected to. Reluctance to have machines running unmanned, then, was not so much the result of union pressure or restrictions, as a position taken by the managers themselves.

In the Japanese factories, on the other hand, one of the attractions of CNC was the potential it offered for unmanned operating. At J1 and J2, CNC machines were left running not only while the owners' sons went in for lunch and dinner, but for several hours in the evenings as well. This is said to be common practice for small factory or *machi koba* owners (Mori, 1982).[4] The families live on the premises, and whoever is in charge of the CNC machines might watch TV in the evenings, and go into the workshop every now and again to check on the machines.

It was not only at the *machi koba*, however, that machines were left running unmanned. It was a conscious goal, and one incentive to purchase CNC for the managing director of J4: 'There were two main reasons why I bought CNC. First, the average age of our operators was 45, and if we didn't get younger operators, we would go under in the end. The old operators were skilled and the young ones weren't, so we had to make up for that shortage of skills, and besides, we wouldn't have been able to attract younger workers without CNC. The second reason was that our machines were only running from 8:00 to 5:00. I wanted to get a few extra hours out of the machines after everyone had gone home and increase productivity.' In actual fact, the operator on the new machining centre was reluctant to leave his machine running after he had gone home. 'I've asked him to do it once or twice a week,' sighed the MD.

At J9, CNC machines were only left running during the lunch hour, with the exception of some wire cutters and spark eroders, which were also left running after work. There were no plans to have the other CNC machines working unmanned. At J45, however, the mini FMS and three FMCs could be left running unattended throughout the night, depending on work loads. At J50, machines were sometimes left running for up to an hour after work, but very small batches were cited as the reason there was not as much unmanned operating as would have been preferred. The story was similar for J66, at which only two CNC machines were left running after work. At J140, experiments with unmanned operating were being carried out to extend machine uptime, and at J180, 40 machines were left running unattended for an average of an hour a day.

In the British factories it was often mentioned that CNC was hard to justify on a shingle shift, and six of the factories actually had a second shift. Of the Japanese factories, however, only J140 and J180 had a second shift, with the occasional exception of the two machines at J66. The managers were reluctant to have more than one shift; 'The biggest reason for not having shiftwork is because the workers are against it. Also, we don't have so many CNC machines, and if we had operators who came for those, there would have to be managers there as well. Then there are problems of rest, of night meals and what have you, so it's not worth doing. We're trying instead to increase our unmanned operating time' (J45, improvements manager). This opinion was echoed elsewhere, although J66 was planning to introduce a second shift with the introduction of FMS, at least in the setting-up period.

That unmanned operating was considered an attractive feature of CNC, however, also depended on a different attitude towards the machines. Part of this difference was in the degree to which the machines were trusted. In the British factories the price of the machines and potential for 'pile-ups' was often mentioned, but references to these were very rare in the Japanese factories. If unmanned operating were not carried out, it was because the batches were considered too small, or it was a goal which was being worked towards but not yet accomplished. Machines were considered obedient objects, and not prone to eccentric or whimsical actions on their own. Human errors – in programming particularly – also caused less concern, despite the fact that the programmers knew less about the machines than their British counterparts. We shall return to these points in the following chapters.

5.2 'Soft' process innovation

If CNC, robots, etc., represent the 'hardware' of process innovation, then QC (quality control) activities, JIT (just-in-time) organization, etc., may be considered the 'software', and it is not uncommon to refer to them in this way in Japan.[5] Considering 'soft' process innovation will further illuminate attitudes towards innovation, which has implications for CNC use.

Small group activities

There is not very much to tell regarding small group activities in the British factories. B71 attempted to initiate QC circles in 1983 but failed. There seemed to have been little attempt to prepare managers or workers for the introduction of the scheme through prior training or education. The industrial relations manager attributed the failure to the fact that 'the workers are all part of the local community, and there is a tradition of long service. Even semi-skilled workers feel responsible for their own work, and they don't like being told by groups or other people what they should be doing in their jobs. It was difficult to set up groups.'

B80 also had a trial of QC circles in 1983 which failed. There was some resistance because workers thought that Japanese working practices were being thrust upon them, but on the whole the personnel manager thought that the manual unions had been supportive and that the problems had arisen mainly at the supervisory level. There was resistance out of fear that supervisors' authority might be undermined, and the supervisors demanded more pay. At the time of the interviews, however, the changeover to modular production was in full swing, and QC-type activities were going to be reintroduced with a different name as part of this changeover, with initial meetings of one hour per week before lunch times. 'We called it education rather than training, and all the module workers received it – what we're making, how and where we are going, right as far as statistical process control. Also, we're not using the words "quality control" this time, but the idea is similar' (personnel manager). This time around, then, there was an attempt to educate workers in QC concepts *before* the introduction of QC circles. This coincided with the abolition of setter/operator distinctions and an attempt to replace individual job identification with module identification to achieve greater flexibility.

B120 had tried to introduce QC circles several times and failed, even with a different name (improvement groups). The unions resisted what they saw as an attempt to force them to do managers' work. None of the other factories had made such attempts.

Suggestion schemes were little more successful. Managers and workers in B39 (and B145) played down the significance of this by saying that top management made a point of walking around the shop-floor frequently, and workers felt they could approach them if they had an idea. Workers' ideas on certain issues were sometimes solicited by the top management as they walked around, too. The suggestion scheme at B71 had been in limbo for the past 18 months, presumably for the same reasons as the QC circles. Greater involvement that managers were seeking was certainly not evident in the suggestion schemes.

All of the larger Japanese factories had QC circle activities, although J66 had only just started them (there had been limited use of ZD – zero defect – concepts before that), with considerable preparation and eventual fanfare. The introduction had been carried out over three years, starting with training of senior managers in TQC (total quality control) concepts. Next were staff responsible for implementation, and then managers in general. This training was largely carried out by consultants from Nikkagiren (Japan Union of Scientists and Engineers – JUSE), and training seminars and weekends were also held for all shop-floor workers to introduce the concepts to them. The first 20 pages of the company's quarterly magazine (for employees) in November 1986 were devoted to TQC and QC circles, starting with the president's address at the commencement of the activities, the union secretary's comments, the vice president's address, speeches from a Nikkagiren consultant, a university professor, and so on.

The speeches emphasized that an important aspect of QC activities is respect for human dignity. Trying to reduce inconsistencies in products does not mean standardizing people as well, said the Nikkagiren consultant. The beginning of problem solving is having each individual bring out the thoughts he or she has within, sharing these with others, and then working together towards the solution of the problems. The congruity of self-development, group sharing and problem solving, and company development – the 'New J66, 70' – was stressed. Even though a lot of money had been spent on preparations for QC activities, however, the reaction from shop-floor workers was not particularly enthusiastic, as the managers themselves recognized. The most enthusiastic comment I heard was: 'Yes,

I'm happy about it. If you work here as long as me, you might as well get into it and give yourself a challenge.'

QC-type activities at J50 go back almost 20 years, with the ZD (zero defect) movement. Nowadays the letters VP (valuable product) are used. As at J66, the unit of organization was the existing work group (*han*). There were still problems regarding communicating between work groups and sections, however, and a large interdepartmental, mutual education programme was being launched. Work groups or sections were to invite speakers from other sections or departments to give presentations on specific topics, including the kind of interdepartmental problems that they hoped for more understanding of. This programme was being launched with only slightly less preparation and fanfare than J66's QC activities.

Small group activities at J45 went under the name of SKY (*shoshudan* – small groups, *kaizen* – improvement, *yakushin* – progress; the company colours were sky blue, also), which started in 1982. An estimated ¥100,000 was spent on training for each 'core' employee (those aged 30 and up) at that time, including managers.

At J180 and J140 attempts had been made to make small group activities fun. The aisles were dotted with QC circle posters, with self-drawn pictures of all the members in the group, group mottoes, scores for various criteria in the past month – preventative maintenance, less scrap, quality, safety, etc. – and also a picture of basketball nets with points scored by each group represented by a different coloured ball, liberally sprinkled with flags. Small group activities were highly evaluated by the personnel managers as raising employee involvement as well as being vehicles for self-development.

What is suggested by this brief passage is that attempts were made to give small group activities several dimensions of meaning beyond that of improving quality and reducing defects. They were to inject variety into relatively monotonous work, provide an avenue for self-expression, and encourage members to interact with each other. From a manager's viewpoint, of course, this would also serve to promote integration. There is a coercive side to this type of activity, since workers who do not cooperate may face censure from colleagues and may also face diminished promotion prospects. The same may be said of suggestion schemes, where workers commonly had a goal (*noruma*, or norm) of two suggestions or so per month. However viewed, and with whatever degree of success, the larger Japanese factories were more committed to this type of 'soft' process innovation than their British counterparts. One might also argue that they achieved more success than their British counterparts because fundamental reforms in employment relations in an organizational direction had already been carried out, which made it easier to present the activities as a convergence of multiple interests.

Just in time

The only Japanese factories which had embarked on just-in-time (JIT) schemes in any systematic fashion were J140 and J180, beginning in the mid 1970s. Along with machine automation, JIT was considered one of the two main pillars to reducing waste (*muda*) through reducing costs, labour, overtime, defects and stocks. At J180 the adoption of JIT practices was blunted by the nature of the small batch work for one-off specialist machines. The machine tools were still largely laid out according to the machine type. The image of the Japanese factory of production lines fed by transfer machines and *andon* lights and *kanban* cards was most closely approximated by the large-batch J140, but it was certainly not how production was organized in most of the factories.

The JIT initiative at B71, as mentioned, had suffered a setback. The new owners required proposals to be broken up into yearly plans, which was not easy to do with JIT proposals, and this and several other projects had been put on hold. B80 and B120 had moved to modular production, similar in some respects to JIT. One interesting feature of the module system, however, and one which was different from Japanese JIT systems, was that the modules were to become accounting centres, and accounting procedures were intended to play an important role in the drive to raise efficiency and reduce scrap and waste. It was still too early for the effects to be seen. Certainly a lot of resources had been mobilized for the changes, perhaps because of all the factories, B80 in particular had felt the full brunt of foreign, particularly Japanese, competition.

5.3 CNC – isolated innovation, or linked?

In many of the Japanese factories there was a widespread belief that conventional machines and traditional skills were on their way out, that they were being replaced by science and scientific techniques. 'When our older operators retire,' said the improvements manager at J45, 'Their skills will be replaced by science.' 'From KKD to NTT' was a slogan at J50, the Ks referring to *kan* (inspiration) and *keiken* (experience), and the D to *dakyo* (compromise); NTT referring to numerical, time and technique.[6] 'From hand work to head work' was the president's new slogan at J66, and so on.

The 'hard' part of this trend, of course, was represented by increasing factory automation, the most obvious of which were CNC machine tools. The 'soft' part consisted of changing skills and training contents, and also QC activities, the intended thrust of which was towards rational problem solving of irregularities in the work process. Both types of innovation were seen in the broader context of a move towards greater application of scientific knowledge and techniques on the shop-floor, and a failure to implement both aspects would result in getting left behind by the competition.

Such concepts as 'from hand to head' and the 'white collarization of blue collar work' were not unknown in the British factories, but were viewed much more sceptically. 'Hard' CNC innovation was thought to require first and foremost conventional machining skills, and these were fostered by giving operators sufficiently long experience on conventional machine tools (see chapter 6). QC circles were viewed as a means of reducing scrap and waste but something which the workers were not likely to take to easily, or rather doubtfully as a means of harnessing worker ideas and commitment, but not as a vehicle for shifting the workers' skill base. Hard and soft innovation were not linked in a larger current of change.

How much, one might ask, were these differing attitudes related to differences in employment relations? This is a question we shall return to in the concluding chapter, but a few observations can be made here. Mention has been made of some of the slogans posted on walls and printed in the publications of some of the larger Japanese factories: 'New J66, 70'; 'New J9 in 10 Years'; 'SKY' and 'AR 2,000' (all round skills etc. by the year 2,000) at J45, and so on. Even the managing director at J4 had a desirable orthodoxy and an undesirable one which he made known to the workers: Orthodox is *o*ptimum, *r*ealize, *t*hink, *h*ealth, *o*bjectives, *d*o, *o*rder, *x* unknown and other things; Orthodox is not *o*bjection, *r*ough, *t*ame, *h*appen (by chance), *o*ld, *d*ull, *o*ut, *x*, unknown and other things.

What is noticeable about these slogans, apart from the penchant for slogans itself, is the constant call for renewal which they convey. Out with the old, in with the new, and in with the scientific. J1 and J2, however, were almost entirely without slogans except for those on safety posters brought in from the outside. There was some visible evidence of them at J4 and J9, and from J45 upwards they were in ample evidence. The unions also had their own slogans, often to do with realizing workplaces worth working in, humanization of work, positive participation of all members to achieve a 'welfare society' and so on.

The greater use of slogans made by the larger factories may be related of course to the fact that in small factories where everyone meets personally every day they are not necessary as a means of communicating ideas, but that they are not an automatic consequence of size may be seen by the fact that the larger British factories made little use of them. The fact that all the large Japanese factories made use of them and the large British factories did not imply national differences beyond the influence of individual factory cultures, but one can also argue that the penchant for slogans is inherent in attempts to foster goal congruence, which is related to OER.

The more organization-oriented a factory, the greater will be the use of slogans which attempt to galvanize the attention and energy of all employees towards common goals and the future. Failure to do this might not only jeopardize company performance, but result in stagnation and flagging morale, and the two would reinforce each other in a vicious circle. Conversely, one might argue that the lack of such slogans and future goals in the small Japanese factories and the British factories is related to the fact that in the absence of supporting employment, payment and industrial relations systems, slogans calling for the realization of common goals and improvements not quantifiable and only remotely linked to higher pay would be greeted with little enthusiasm. Although these systems were undergoing change in most of the British factories, many of the changes were reactive rather than proactive, designed to reduce past tensions. The company was still first and foremost there to produce products, to make a profit in doing so, and possibly to support the livelihoods of the workers. Notions of self-actualization and group dynamics were being toyed with – at B80 by personnel managers, for example – but were not integrated into company goals, institutions and activities. They were not infused into QC circle-type activities and routine work activities as they were in the larger Japanese factories.

There is thus a sense in which employment relations can be said to exert a direct influence on attitudes towards change and innovation. This is not the whole story, though. Attitudes towards unmanned operating, for instance, were influenced by other factors. I shall return to these as well as to the influence of employment relations in the following chapters as I examine training of operators for CNC, the division of tasks around CNC, and CNC skills.

6

Training

Attitudes towards CNC offer some clue as to approaches to operator training. Machines which were unpredictable required highly trained operators. Predictable machines required technical knowledge. The country differences were striking, but there were also differences within the countries. These were not necessarily in the direction predicted by the hypotheses. Just to remind the reader, those hypotheses were:

1 More training is given to CNC operators where employment relations are organization-oriented than where they are market-oriented.
2 A wider range of tasks is performed by CNC operators where employment relations are organization-oriented than where they are market-oriented.
3 The influence of factory size and batch size on CNC use is mediated by employment relations.
4 Skill levels are preserved or enhanced where employment relations are organization-oriented and contested where they are market-oriented.

We are concerned mainly with the first hypothesis in this chapter. Training is fundamental to any discussion of skills, yet surprisingly is often left out of discussions of 'deskilling' (or 'reskilling') and new technology, which tend to focus instead on task ranges and discretion. Training is divided into general or career training and training specifically for CNC operating. The key question asked in the first section is how much and what kind of experience operators were given on manual (conventional) machine tools before they were assigned to CNC. This will be the measure of career or general training, but before looking at this, let us look at the educational backgrounds of operators, and different approaches to training, particularly craft and on-the-job (OJT) training. The second section looks at CNC-specific training, both on and off the job, in-company and out-of-company, and finally the factories are evaluated, ranked and the results compared with the OER–MER rankings.

Training

6.1 General training

Education backgrounds

Although the first hypothesis is concerned with *employee* career development and training, general education undoubtedly does influence this. The Japanese education system is at present a subject of interest in several western countries, Britain included, and a number of studies suggest that a greater 'three R' competence is attained by a greater proportion of pupils there than in Britain (see Prais, 1987). If a higher level of 'three R' competence is indeed attained in Japanese schools, one might predict that operators in Japan would acquire CNC-related skills more quickly than their British counterparts, particularly mathematics-related skills, and there might be more learning from manuals. There may, therefore, be different approaches to training. Dore and Sako (1987), for example, argue that Japan's basic education system produces people capable of following carefully detailed and complex written instructions. A lot of learning is based on informal production of job specifications and procedure manuals meticulously written out by supervisors and used as self-teaching material for newcomers in a job.[1]

Most of the Japanese CNC operators in this study had completed 12 years of schooling from the age of 6 to 17 or 18; for some of the older ones in the larger firms this meant nine years of schooling and a further three years in the company school, while some of the J2 operators were also attending evening high school classes. Most of the Japanese factories preferred to recruit employees from vocational high schools, in which, however, the curriculum is heavily loaded with general subjects (Ishikawa, 1987). There were two reasons for this. First, hiring them for their knowledge of machines, practical and theoretical, was of some importance, although some foremen and managers expressed reservations about the level of technical skills and knowledge actually gained. Secondly, to some extent these schools also socialized their students to accept the kind of jobs they would likely be given in the factories: 'blue-collar' jobs.

As mentioned in chapter 3, however, J140/180 preferred to recruit from general high schools because they wanted to 'mould' the employees themselves and because they wanted to recruit school leavers further up the ability range. In general, general high schools take in and educate pupils higher up the ability range than technical high schools, although there is some overlap. There are also rankings within the general high schools, and within the technical high schools, which employers are well aware of. Larger employers (including J140/180), which are famous, pay higher wages, etc., are able to recruit workers from higher up the ladder, hence their operators will have a higher level of scholastic achievement than those in the small factories.

The British operators were also not in the upper range of their age group in terms of mathematical skills: 'Someone with 6 O levels will go on and do A levels and go into an office or a technical job. A craft apprentice won't have O level maths' (B80). Even though B145 was getting 20 times the number of applicants it needed for apprenticeships, it was cautious about raising the academic requirements too high (presumably because the trainees would soon want to leave the shop-floor). The industrial relations manager of B71, however, said that his company did require O levels in maths, English and science.

Whereas the Japanese operators had received 12 years of school education, the British operators had received a maximum of 11, with fewer hours per year. Furthermore, as Dore and Sako (1987, vii) argue: 'The most striking British–Japanese difference in three-Rs attainments is among those in the bottom half of the ability range.' This means that the overall range of scholastic abilities of operators in the factories was probably large, possibly comparing with the range of scores on the OER–MER continuum given in figure 4.1, except that of the British factories, the larger ones were able to be more selective, at least regarding apprentices. These differences might affect not only approaches to training, but the degree to which managers consider it worthwhile investing in training at all.

Differences in starting points, then, should be kept in mind in the discussion of operator training, particularly for example when we see that a number of Japanese operators taught themselves to operate CNC machines from handbooks alone.

Britain and craft training

Managers look for a number of qualities in those they select to operate CNC machines, including the following:

1 *general intelligence or 'brightness'*: the ability to absorb new ideas quickly and act according to well-reasoned, logical patterns of thought;
2 *sense of responsibility*: conscientiousness or diligence, doing one's best in an unfamiliar or familiar situation;
3 *machining skills and knowledge*: knowledge about metals, tools, setting, and experience with particular machines.

Most managers will want a combination of all three of these, but may differ in emphasis. There were in fact differences within the British and Japanese factories and between them. In the British factories, relatively heavy emphasis was placed upon (3) machining skills and knowledge, while in the Japanese factories more emphasis was placed upon (1) general intelligence or 'brightness'. All managers looked for (2) sense of responsibility, but in the Japanese factories, particularly the large ones, it was assumed that operators would have this, whereas in the British factories it was sometimes explicitly mentioned as a quality that managers were looking for.

A number of people have written in recent years on the waning influence of craft training in Britain. The number of apprentices has decreased six-fold in the past 30 years (Brown, 1986, 162), and training is supposedly becoming more firm specific. In 1987, B145 and B71 were training about one third of the apprentices they were in 1967. The number of employees at B145, however, had more than halved over the same period of time, and at B71 they had been reduced by more than 30%. At the same time, the retention rates of apprentices had increased in both – by more than 300% in the case of B71. While indeed craft training in Britain might be declining, it will be argued here that it was still very important in both unionized and non-unionized factories.[2]

In fact, in some of the larger factories the preference for time-served operators had increased. We saw in chapter 5 that B145 managers were tempted to use semi-skilled operators on NC at first, but had changed their minds. With two exceptions the non time-served operators were all working on machines classed as 'semi-skilled'.[3] 'They can see problems on the horizon. And anyway, our supervision in here is so sparse that they need to be skilled' (superintendent). At B71, all the CNC operators were time-served. The IR manager summed up the general (present) consensus; when asked what the most important requirement of a CNC operator was, he immediately replied, 'Anyone who has come through the formal training (apprenticeship) should be good enough.'

At B120 the six non time-served operators were operators before 1984 when the setting tasks were done by specialist setters, and were retained when the setter/operator distinction was abolished. All subsequent operators have been time-served. The same might soon become the case at B80 with the imminent abolition of the setter/operator distinction there according to a training manager. In fact, the strengthening of craft preference had caused a *reduction* in internal promotion: 'After the war there was a kind of progression from operators to setters to supervisor. With the type of machine we have nowadays – including CNC – we want people with a bit more savvy, who have had proper apprenticeship training' (personnel manager).

Concurrent with the strong demand for craftsmen at these factories, more and more upgrading training which was considered firm-specific was also being given. This suggests that an increase in firm-specific training does not necessarily coincide with a decline in craft training. There is, of

Table 6.1 *Time-served CNC operators (British factories)*

Factory	No. of regular NC/CNC operators	Proportion time-served (%)
B4	3	67
B8	5	80
B11	3	50[a]
B12*	23	26
B39	11	36
B71	26	100
B145	70	80
B80*	30	0
B120	38	84

*Indicates large batch factories.
[a]One operator had almost finished his apprenticeship before the company went bankrupt.

course, a certain degree of firm specificity to apprenticeships themselves. While B120 ran a four-year apprenticeship – teaching general principles in the first year then giving apprentices three months on different machines and in different departments including assembly and inspection with specialization in the fourth year – B71 ran a three-year apprenticeship, with apprentices getting most of their experience on different machines in the first year and specializing in the second and third. Moreover, apprentices could finish in less than three years if they demonstrated the appropriate level of skills.

What was it about apprenticeships that the managers in these factories found desirable when selecting CNC operators? It could be general intelligence (being 'bright' enough to be selected for an apprenticeship in the first place), a sense of responsibility (being responsible enough to last the course, and having 'rough edges' taken off in the process) or machining skills and knowledge (theoretical knowledge gained from the off-the-job part of the course and skills gained through working on different machines). Of these, the last was the most important. The craftsmen could 'see problems on the horizon', 'tell from the sound of the machine if everything was all right' and 'know how to make the machines work'.

It may be possible to gain these skills through machining experience without an apprenticeship. Some managers and supervisors in the larger factories agreed, but thought this would be exceptional, while their counterparts in the smaller factories were more equivocal. At B39 the production manager said there was a 'tradition of internal promotion' and that he preferred internally promoted unskilled and semi-skilled workers for his

CNC operators: 'If we train a truck driver or a sawyer to be an operator, and finally work on CNC, he is much more likely to stay than a craftsman operator. A craftsman might be tempted out into tool-making shops for a few extra shillings, but the others wouldn't be so sure of their skills to go out into different work with [piecework] incentives.' He kept a look out for potential CNC operators for several years in advance, and the sawyer manager, for example, was very happy when his workers moved up into operating. However, the company did advertise for craftsmen, did train them, and would have had more on CNC if it were not so hard to recruit or keep them. In other words, B39 was less insistent on having craftsman operators, partly because of the difficulties in training and keeping them. It was also less insistent on machining experience itself.

At B12, too, the managing director declared there was a policy of promoting unskilled and semi-skilled workers on to CNC – those who demonstrated some initiative. The time-served operators plus two others, however, were an elite on the shop-floor. They were the only ones trusted to do setting up by themselves. The foreman at B11, who championed the introduction of CNC and selected operators, also declared: 'Anyone who comes in here with a bit of initiative can be on top in two years', which is precisely what had happened in the case of one operator. On the other hand, the only non time-served CNC operators at B8 and B4 had been assigned to CNC drilling (which was comparatively easy) to replace operators who had left. While B4 was the only factory which did not do apprenticeship training, it recruited time-served operators where possible. The problem in the smaller factories was keeping good craftsmen: 'Let's face it, the able ones tend to move on and those who can't stay here' (foreman, B11).

Japan and OJT

Apprenticeships and internal promotion/on-job-training (OJT) are sometimes viewed as dichotomous modes of training (e.g. Koike, 1977; the words of the manager at B39 also suggest such a dichotomy). Craft training allegedly takes the place of internal promotion-OJT, encourages a 'once-and-for-all training' mentality, and discourages movement – or job rotation – on the shop-floor. There is some truth to this argument, and we will return to it at the end of this section, but some reservations must also be expressed. First, as argued above, any increase in the firm specificity of training has not necessarily weakened the value managers attach to apprenticeship training or apprentice-trained workers. Secondly, interviews with managers and CNC operators in both countries suggested that there was not, in fact, more mobility on the part of the Japanese operators, although the observation cannot necessarily be generalized to all shop-floor workers, since CNC operators in Britain were normally the 'cream', and as such were mobile. A number of British operators thought that there was more mobility now than twenty or even ten years ago (which also applies to Japan – Mihara, 1986).[4] Thirdly, as we are about to see, the absence of apprenticeship may encourage but does not ensure planned internal rotation and OJT.

Apprenticeships in Japan disappeared after World War II, and it is generally considered that 'craft' mentality is now confined to older workers in small factories (who identify more with their craft than their company, and who have a craft 'pride').[5] The main form of training in Japan is OJT, with workers supposedly acquiring skills through performing jobs of gradually increasing difficulty, one job being the foundation for the next.[6] There *are* super-corporate qualifications in the form of certificates that workers may obtain by passing *gino kentei* (craft or skill examinations), which were set up by the government in 1959, and by 1984 taken by almost 3 million workers (with a pass rate of 42.6%) in 128 trades. While many companies encourage workers to take the tests, the certificates are not normally given much weight in job changing, except by smaller companies, nor are they a sufficient condition in themselves for promotion (Ishikawa, 1987, 18).[7]

All of the larger factories encouraged their employees to take the skill tests. J140/180, in fact, administered its own, which paralleled the government tests but which were held to be even more difficult. Taking these tests was semi-compulsory. The criteria laid out for the tests are meant to serve as criteria for study and self-improvement. Three years after entering the company, employees took grade 2 tests – for lathe work, milling, drilling, and also for office and non-production work – and after a further two years they took the grade 1 tests. The system was developed because: 'You can't keep up with rapid technological innovation if workers just do as they're told. You need something to draw out their enthusiasm, and we were very concerned about developing the abilities of the workers' (personnel manager).

This did not mean, however, that all operators selected for CNC work had gone through the school, done the skill tests, or even worked from simpler manual machines through progressively more difficult ones and on to CNC in a systematic fashion (as in the aircraft factory of figure 1.1). At J140, workers were taken on to CNC after half a day's instruction of pushing buttons, putting in tapes and checking offsets and quality. Little movement was reported between different types of machines, such as mills and lathes, because of 'production needs'. Even at the elite J180, about 10% of the operators had been taken directly onto CNC: 'Ideally they should have manual experience, but we just can't keep up.' This suggests that OJT was not as structured as it might have been, despite clearly laid out promotion criteria. More importantly, it suggests that relative to J140/180's British counterparts (B80 and B120), less machining experience was considered necessary to begin CNC work, even if this was to be acquired in the course of time.

J66 gave its operators the most experience on manual machines before assigning them to CNC. OJT was interspersed with periods of off-JT, but even there the personnel manager lamented that the OJT part was not as well organized as it should have been. J50 was probably the most systematic; the following shows the sequence of jobs done by two operators prior to being assigned to CNC.

Operator A
sub-operator, plane/mill 1 year
operator, small lathe 2 years
operator, vertical lathe 1 year
operator, facing lathe 2½ years
operator, jig borer 2½ years
preventative maintenance 2½ years
operator, CNC borer

Operator B
deburring 6 months
sub-operator, borer 9 months
operator, small lathe 2½ years
sub-operator, CNC borer 1 month
operator, CNC machine centre

It may be significant that J50 had the least number of CNC machines apart from J1. In the smaller factories the backgrounds of CNC operators suggest that training was often a matter of 'satisficing'. Operators were familiarized with tools and basic machining operations during their probationary *three months*, and were then taken directly onto CNC if they proved adaptable enough (whereas the training/trial period in J50 and J66 was in principle about three *years*, when the operators took the grade 2 skill tests). J9 was the most thorough; three trainee operators were put under the supervision of a leader for six months or so beyond their probation period to deepen their skills. In J2 and J1, operators were taken directly onto CNC without even a probationary period.

In summary, job rotation, careers and OJT were not as systematically implemented as personnel policies might stipulate. This does not mean that little training took place in the factories or that managers were not committed to training their employees. As argued in chapter 3, the ideal and intent were present in the larger factories, but they were shaped by time horizons and 'production needs'. Secondly, the above suggests that many Japanese managers considered CNC operating not to need the same degree of machining skills and knowledge that most British managers did. Many, in fact, were willing to take workers directly onto CNC without experience on manual machines at all. Let us now look at this more explicitly.

Experience on manual machines

In no less than five of the Japanese factories had workers been taken directly onto CNC machines without prior experience on manual machines, and in two others with six months or less experience (table 6.2). In J140, the whole question of prior manual machine experience was considered meaningless because 'those skills are not needed now' (production manager). Almost all new operators are now taken directly on to CNC, while at J140's British counterpart B80, experience on manual machines was considered very necessary, and increasingly so. In one of the British factories – B12 – four workers had been taken directly on to CNC to do simple operating work ('button pushing') but it was decided that this arrangement did not work well, and it would not be done again. B39 had sent two workers with no prior machining experience to a skill centre for six months followed by a CNC course, but had decided that more machining experience was preferable.

There were differences within the countries as well as between them. The Japanese factories can be divided into four groups:

1 **Smallest factories**: Workers taken directly onto CNC with no machining background. Manual machine skills considered unnecessary.

At J1 an operator with some experience on manual machines would have been preferred, not so much for machining skills but: 'If they go straight onto CNC they don't know the worth of it. They don't know how accurate it is, and how it saves on heavier things (i.e., less moving and lifting).' The new recruit had previously worked in a shoe shop. At J2, a 17-year-old had worked on a bench miller and done odd jobs for a year before being put on to CNC, but the other two had no experience in a machine shop beforehand. The brother in charge of CNC maintained: 'CNC doesn't require a great deal of skill, and information about speeds and feeds can be found in the handbook.'[8] Having young and inexperienced operators was partly related to keeping wages down, but the practice reveals a certain attitude towards CNC; that it replaced conventional skills rather than building on them. In both factories the manual machine workshop and the CNC workshop were separated, with little interaction between them.

2 **Smaller factories**: Workers put on CNC after probationary period. Manual machining skills considered necessary eventually to supplement CNC skills. CNC at J4 was intended by the MD to make up for the loss of conventional skills as the older workers retired and more and more school leavers were taken on. Although they came from technical high schools, their knowledge of machining was very rudimentary, according to the foreman. At J9 the need for manual machine skills was stressed more, but

Table 6.2 *Prior experience on manual machines (years)*

Factory	Minimum	Maximum	Average	Factory	Minimum	Maximum	Average
J1	0	6	3	B4	4	38	15
J2*	0	4	1	B8	6	37	22
J4	0.4	2	1	B11	2	20	11
J9	0.4	27	3	B12*	0	8+[a]	4
J45	0	12	6	B39	0.5	20	6
J50	3.5	9	7	B71	3	23	9
J66	6	29	14	B145	3	30	7
J140*	0	—([b])	—	B80*	3	35+	10+
J180	0	5	2	B120	3	35+	15

Note: British figures do not include first year of apprenticeship.
*Indicates large batch factories.
[a]'Eight years in the company, probably more.'
[b]'That question is irrelevant because those skills are not needed now.'

rapid expansion had meant that they could not take experienced manual operators on to CNC: 'Ideally, they should have a lot of experience on manual machines. At the very least, they have to know about tools, so we try to give them two or three months to learn about those before we put them onto CNC... In half a year, they're half fledged.' At both factories manual machining experience was to supplement CNC work, as young operators discovered their limitations and wished to experiment on manual machines themselves.

3 **Medium-sized factories**: Manual machine skills a prerequisite for CNC work. Machining skills and knowledge stressed.

The emphasis on manual skills in the medium-sized factories was probably more a product of their industries than their size; J50 and J66 belonged to what is generally considered a conservative industry in that customers prefer tried and trusted engineering principles, and product innovation often consists of applying more robust materials, etc., rather than radical design and other high-profile changes. Also: 'A lot of our work is in stainless, and the batches are very small, so our operators have to be reasonably skilled' (J50). J50 and J66 were also the only Japanese factories apart from J1 which saw a decline in the number of employees from 1977 to 1987 (table 1.4). Growth and change in their industry were not as much a facet of everyday life as they were for factories like J9 ('Life is innovation; some of the workers are worn out from it,' said the MD) and large dynamic ones like J140/180. J45 had until recently made machines (of high engineering content) for a government monopoly, which may also have fostered a (less) conservative view of skill requirements: 'At first we had very skilled operators working on them (NC/CNC), now not so skilled. We've got a combination now' (improvements manager). At J50, too, however, it was envisaged that younger operators with less experience on manual machines would be used in the future.

4 **Large factories**: General intelligence and adaptability stressed. Manual machining skills a part of skill testing however.

Finally, there were the factories J140 and J180, which belonged to an industry at the forefront of production technology. The rapid growth of the company and large amount of capital investment have also served to foster a 'culture' of change. The workforce was very young, and workers were supposed to learn manual machining skills through self-education or with lunchtime guidance in many cases, while job assignments were influenced by fast growth and associated production needs.

There were also differences within the British factories, as noted:

1 **Smaller factories**: Mixed attitudes towards craft, but manual machining skills a prerequisite for CNC operating.

In the smaller factories, B39 included, managers were more apt at least to talk about their 'internal promotion' and how anyone with the right attitude and aptitude could be promoted to CNC work. The proposition that (1) much machining experience is necessary for CNC operators and that (2) an apprenticeship ensures that the operator has these were open to question. This was related in part, however, to the availability of craftsmen, and at B8 and B4, where experienced operators were clearly preferred, the 'skilled = time served' equation was still strong.

2 **Larger factories**: Craft training emphasized, often a prerequisite to becoming a CNC operator. Machining skills still a prerequisite in other cases.

For the managers and supervisors in the large factories, apprenticeships gave operators a wider understanding of machines, tooling and potential breakdowns than could be gained by on-the-job work experience alone, and their own craft backgrounds led them to accept both the propositions above (if they were conceptually separated at all). Thus the decision to use 'skilled' (= time-served) operators was not simply for the sake of industrial harmony, although such considerations no doubt played a part. Remarked the convenor at B145 in reference to two upgraded operators: 'It's all right with us provided they have the skills for the job and are paid the skilled rate. Anyway, management want skilled operators.' Some managers did think that much of the CNC work was routine, and that craftsmen would eventually get bored working on it, but they preferred to have them there in case something went wrong – the uncertainty factor referred to in the last chapter. Even the work study manager at B120, who thought most CNC work was quite easy, thought that operators should not only be craft-trained, but have one or two years of experience on manual machines afterwards as well.

There are other possible ways of dividing the factories up. Batch sizes did influence views of skill. At J140, manual machining skills were considered unnecessary, and J2 also took workers directly on to CNC. There were relatively few craftsman operators at B12 and none at B80, but the operators there still had more experience on manual machines than those in several of the small-batch Japanese factories.

If anything, there had been a trend over time in the British factories to use more experienced operators as the 'trained monkey' view lost its credibility, while in the Japanese factories the reverse was the case. Even J140 originally had experienced operators: 'When we first introduced NC we had special operators. With changing times and the computer age, new workers, even the ones with no machining experience, have no resistance to CNC, so we can have not very skilled operators turning out precision parts.'

Age

One thing which has an important bearing on 'general intelligence and adaptability', 'sense of responsibility' and 'machining skills and knowledge' is age. Table 6.3 shows the averages and range of operators' ages. In none of the Japanese factories was the average age greater than 35, while in three of the British factories it was over 40. In five of the Japanese factories the average age was less than 30, while this was the case in only one British factory. There were regular operators less than 20 years old in four of the Japanese factories, but in none of the British factories.

These differences are in part attributable to overall workforce age differences – the average age of conventional operators was also less in many of the Japanese factories – generally less than 40 in Japan and over 35–40 in Britain (table 6.4). Overall age profiles, however, only partially account for the differences. In the Japanese factories, only one operator was assigned to CNC after the age of 35 (at J66), whereas a number in the British factories had been assigned after the age of 50.[9] CNC in Japan was said to be a young man's job. Older operators were supposed to be reluctant to go on to CNC – they had an 'NC allergy'. Not only was this view widespread, it seemed to form an *assumption.* Older workers had not been tried out on CNC, they were normally not considered. (In J1, for example, when the son of the owner was trying to find a new operator for CNC, and was struggling to

Table 6.3 *Ages of CNC operators*

Factory	No. of operators	Youngest	Oldest	Average	Factory	No. of operators	Youngest	Oldest	Average
J1	2	20	35	28	B4	3	20	56	35
J2*	4	17	36	27	B8	5	24	55	41
J4	5	18	22	20	B11	3	29	42	34
J9	15	21	48	35	B12*	23	21	64	30
J45	13	23	41	34	B39	11	23	42	31
J50	3	23	40	31	B71	23	23	46	30
J66	8	24	47	32	B145	60	21	51	28
J140*	300	18	35	23	B80*	30	22	62	42
J180	100	19	35	27	B120	38	21	63	41

*Indicates large batch factories.

operate both machines by himself for several months, he did not think of asking the older operators in the manual machine shop to help.) This attitude was strongest in the smallest factories – which were most concerned about labour costs, and younger workers would mean lower labour costs – but managers in all factories mentioned the age factor and often the 'NC allergy' of older workers. In most of the factories the concern was over 'adaptability', which young workers were supposed to have and older workers were not.

In some of the British factories, too, youth was important in the selection of operators. At B39 both managers and operators thought that age was a significant factor in being able to master CNC, and in the selection of operators. At B145 the ideal age of operators was held to be in the late 20s, which was the preferred age in some of the Japanese factories (although younger in some cases).[10] The production manager at B8 and the managing director at B4 would have had younger operators on CNC if they had been available, but they weren't so older operators were trained. Older operators, however, were not excluded upon an *assumption*. It was after training and subsequent difficulties that it was concluded that younger operators were preferable. Some older operators were confused by talk of binary operations and programming during training courses. Even if older operators could not master programming, though, they were considered quite capable of operating the machines.

At B71 and B80 an age factor was specifically denied (although in both cases the average age of CNC operators was less than that of manual operators): 'No, there is no age factor. Experience and ability come first' (B80). 'Anyone with the skills should be qualified to do it, no matter how old

Table 6.4 *Ages of CNC and manual operators*

Factory	CNC	Manual	Factory	CNC	Manual
J1	28	42	B4	35	55
J2*	27	40	B8	41	49
J4	20	47	B11	34	40
J9	35	28	B12*	30	38
J45	34	37	B39	31	38
J50	31	40	B71	30	35
J66	32	39	B145	28	45
J140*	23	26	B80*	42	48
J180	27	28	B120	41	42

*Indicates large batch factories.

they are.' 'No, we don't do anything like that here' (referring to the deliberate selection of younger operators; B71). This view reflected an egalitarian conviction, that just like height or dialect, age in itself was irrelevant if the person had the 'skills'.

In sum, age was largely related to experience in the British factories, and sometimes to adaptability (and responsibility), but the reverse was the case in Japan.

Youth versus experience

Several reasons may be given for this relative youth and lack of experience on the part of Japanese operators. First, there were differences in education backgrounds, which we looked at briefly at the beginning of the chapter. It could be that with their education the Japanese operators were able to learn how to operate CNC with less experience on manual machines than the British operators.

Secondly, supervision was more intense in the Japanese factories, which might make up for a lack of skills and/or provide for more intensive OJT. The sparseness of supervision at one British factory (B145) was specifically cited as a reason why experienced operators were needed, while at one Japanese factory (J9) closeness of supervision was cited as a reason why it was possible to put operators with little experience on manual machines on to CNC.

In the smallest factories, the family member in charge supervised one (J1) and three (J2) other operators. In J4 the ratio was 1:5, at J9 there was one leader for every 1–3 trainees, and otherwise one group leader for five operators. In the larger factories supervision was also close. At J45, for example, there were about 30 operators under a foreman, but they were divided into 4 groups with a group leader in each. This group leader decided who should do what work, and was responsible for training in the group. There was, moreover, an *assistant* group leader. These different levels formed a hierarchy not so much of formal authority – although they were by no means inconsequential in this – but of promotion, skills and also of training. The same applied to the other large factories.[11]

In the smaller British factories, with few operators, supervision was naturally close. At B4 there was on average one foreman for 8 operators, but no leading hands under the foreman. At B8 there were 12 operators under one foreman, and one of these was a leading hand. From B11 up, the ratio tended to be about 1:15, although in the machine shop at B71 (they were overmanned on supervision, according to some managers) it was 1:10. Only the semi-skilled sections of the larger factories had chargehands, with the exception of B145. Here, however, as at B71, chargehands were being phased out as part of an effort to give operators greater responsibility for their work.

Of course there were also operators in the British factories who were recognized as being particularly skilful and who sometimes gave informal instruction. Operators on machines similar to ones about to be purchased also gave instruction. The closer supervision in the Japanese factories, however, was intended to serve as a training mechanism for operators with little prior experience, while one can argue that apprentice training in the British factories both obviated the need for closer supervision and also precluded it.

Apprenticeships, then, were influential not only as a means of training and in the selection of operators, but also in the organization of workers. It did seem that the craft backgrounds of the British managers led them to view CNC as a technique requiring of operators first and foremost machining experience if not a craft qualification, whereas the absence of these in Japan led managers to view CNC as requiring first and foremost adaptable operators, or as some put it, operators with 'a bit of maths ability'.

6.2 Training for CNC

We now go on to look at training specifically for CNC, and will find that those managers who emphasized the importance of manual machining skills also emphasized CNC-specific training. In table 6.5 this has been divided into six categories: attendance at skill centre/polytechnic courses,

Table 6.5 *CNC-specific training*

Factory	No. of operators	At skill centre	At manufacturers	Off-JT in company	On-JT (by manufacturer)	On-JT (by employee)	Mainly from handbooks
J1	2	0	0	0	1	1	1
J2*	4	0	1	2	2	3	1
J4	5	0	5	0	0	2	0
J9	15	0	5	0	7	8	0
J45	15	0	5	0	'some'	5	3
J50	3	0	3	0	3	0	0
J66	8	0	0	'some'	'most'	'some'	0
J140*	300	0	0	260	200	100	0
J180	100	'rarely'	50	90	50	100	0
B4	3	1	0	0	0	3	0
B8	5	0	5	0	0	4	0
B11	3	1	2	0	0	1	0
B12*	23	6	3	0	10	13	0
B39	11	3	4	0	2	8	0
B71	26	0	2	0	16	16	0
B145	70	30	3	70	'many'	69	1
B80*	30	12	2	15	3	'most'	0
B120	38	'most'	6	'most'	'rarely'	'most'	0

*Indicates large batch factories.

courses at the manufacturers or vendors, in company off-job training, on-job training from the manufacturer's engineer, on-job training from someone in the factory or company, and learning mainly from handbooks.

Outside courses (at skill centre or manufacturers)

One of the striking things about the table is that almost none of the Japanese operators had gone on courses at skill centres or polytechnics, whereas a considerable number of British operators had, and the figures would have been much higher had manual machine operators been included. Furthermore, more operators might have gone (e.g. from B71) if the machine tools at the local polytechnic had been more up to date. While such centres and courses did exist in Japan, the common attitude was why send operators out on courses when they can be trained in the company.[12]

This lends support to the argument that where external labour markets are predominant, upwardly-mobile workers must seek training outside the company at their own expense (i.e., buy the skills they will sell). The argument is compromised, however, by the fact that very few of the operators had gone on outside courses at their own expense. One had in B4, one in B11, and two or three in B120, but the vast majority had gone at company expense. This was contrary to what was anticipated in chapter 1.

In selecting workers to go on such courses, managers were keen to be seen to be fair, or concerned about their workers, and participation was generally open to all operators, although there was a selection of those chosen to go on to subsequent courses based on test results at the end of the first course. B80 had sponsored a specially designed course for many of its setters and some of its operators. The courses ranged from two week block courses to 2–3 month day release courses to 6–12 month evening courses.

This does not mean that the Japanese factories spurned all external courses. Like the British factories, they widely used the training courses that came as part of the CNC purchase package. Courses run by manufacturers in Japan tended to be three days or sometimes four and occasionally longer, whereas in Britain they usually lasted for a week. Again, the figures above show CNC operators only, while programmers and sometimes supervisors and maintenance workers were sent in both countries. Three people were often sent in the British factories, however, while the Japanese factories sometimes sent only one – a production engineer when the technology was new, or a group leader where it was not a major introduction – who would come back and pass on what he had learnt to others in the company. At J66, for example, production engineers first went on courses and taught NC/CNC techniques to the operators when they returned. Now for individual machines it is the group leaders in the machine shop who go. This process was not unique to the Japanese factories; at B120 it was mostly production engineers who were sent on the courses, and they trained the operators. At B71 in-house training had become the preferred means of training operators, too.

There did seem to be a greater reluctance to upset production routines in the Japanese factories for this kind of training – which might explain why the manufacturers' courses were shorter – not only in the factories with the most CNC expertise, but also in the smallest factories. 'I went on a course for our second CNC but soon stopped. When you're working for yourself you can't afford to waste the time' (J1). The new operator was being taught on the job. At J2 the factory manager was thinking of sending the operators on a course, but it would have to be in their summer holidays.

On the other hand, when J50 purchased its first CNC machine, three workers were sent to the maker's factory (which was part of the industrial 'group' J50 belonged to) for a month, not only to learn about CNC, but also to observe how CNC was used there. Like the British managers, J50 managers not only required CNC operators to have a lot of experience on manual machines, they also considered relatively lengthy CNC training to be necessary.[13]

Off-job training (in-house)

The largest Japanese and British factories carried out in-house off-job CNC training (J2 is also shown as having done some; two operators were given mainly mathematics instruction while longer batches were being machined automatically, but this was related to programming). At J66, upgrading training was carried out every 2–3 years, and this included CNC skills. The training consisted of weekend or afternoon–evening courses. There were five CNC machines in the J140/180 company school. Those students doing the mechnical course spent two weeks working on them, while those in the electrical course spent one week. Workers already doing production work were also sent in for training. This was sometimes done instead of OJT at J180.

At J180's British counterpart B120, apprentices also spent at least a week learning CNC in the training centre, and some subsequent CNC training was done there. Many setters and operators at B80 had studied CNC at the factory's training centre as part of a six-evening and two-Saturday-morning course. B145 had run short orientation courses for its operators to 'try to take the mystique out of CNC'. Some of those operators were very reluctant to go on courses outside the company because they feared being put into a competitive environment, and it was alleged that the 'colleges talk in binary language, and what they teach is out of date, anyway' (the first complaint was also often levelled at manufacturers' courses).

On-job training

OJT in table 6.5 is divided into OJT under the supervision of an installation engineer, and OJT under the supervision of someone else in the company. When CNC machine tools were purchased, an engineer from the manufacturer generally came in to ensure that the machine was successfully installed and running. Unfortunately, some of those interviewed regarded this as a kind of training while others did not (in both countries), so it was difficult to obtain reliable comparative figures.

While operators often learned from the manufacturer's engineer at installation time, when they were assigned to CNC at other times they almost always learned from the foreman, chargehand or the incumbent. Sending operators on courses which were not part of the purchase package was considered costly (at B8, for example, which paid £500 for each of two operators for a course plus hotel and other expenses) and was avoided. If anything, the British foremen claimed that more of their operators were trained on the job by someone in the factory and fewer by installation engineers than their Japanese counterparts. This was partially attributable to a greater turnover of CNC operators – promoted (they were the cream) or leaving – which necessitated training at a time other than purchase time. It was also attributable to greater numbers of night-shift operators, who often had learnt from co-workers.

Learning from handbooks

As might be expected from the preceding discussion, with one exception those who taught themselves mainly from handbooks (the word 'manuals' is avoided here to prevent confusion with manual machines) were Japanese. The exception was a senior operator at B145, who was a computer buff. He had largely taught himself, and was now considered an expert and used as a 'trouble shooter'. The figures in table 6.5 possibly underestimate the amount of self-teaching in both countries, but particularly in Japan. Other operators, for example at J4, had spent three days on a CNC course (with little experience on manual machines) and had to pick up the rest by themselves when they came back because their foreman had no knowledge of CNC. At both J4 and J9 operators reported having to take stomach medicine at first because they were 'thrown in at the deep end'. One operator at J45 reported a similar experience when a CNC operator broke his leg and he had to take over his machine: 'It was frightening at first, trying to get the machine going while reading the handbook. At first everything I did was wrong, but I mastered it in six months.' Even where OJT was given, self-instruction was expected; 'Speeds and feeds? Oh they can look those up in the manual' (at J2, but echoed elsewhere). Book learning also took place in some of the British factories: 'Our star operator was a welder. He took home some books and went on some courses, and when the machining centre came in, he was the first to be called' (B12). It was said to be the *main* means of learning for only one operator, however.

It may be self-learning from handbooks in the Japanese factories took the place of sending workers on outside courses at polytechnics or skill centres (or vice versa; that these courses took the place of book learning for British operators). It was clear, however, that CNC was perceived by the British managers as requiring not only more experienced operators, but also more CNC-specific training. While young, adaptable operators with a bit of mathematical sense were supposed to acquire CNC operating skills in the Japanese factories, more experienced operators with longer OJT times and often outside training were thought necessary in the British factories. 'Trial and error' was not an acceptable way of gaining expertise as it was at J45 (above) and J1: 'I give him a chance to do things by himself until he runs into a problem. That's the only real way to learn.'

6.3 Training, OER and MER

The first hypothesis stated:

More training is given to CNC operators where employment relations are organization-oriented than where they are market-oriented.

Training scores

In chapter 4 the 18 factories were rated according to OER–MER. The same will now be done for training. Training is divided into general or career training and CNC-specific training, and for scoring use is made of tables 6.2 and 6.5. The method of scoring is as follows. For general training, experience on manual machines has been taken as the indicator. The factories have been given a score from 0 to 10 depending on the average number of years operators were given on manual machines before starting on CNC. These figures correspond to the average number of years of table 6.2, with an upper ceiling of 10, the subjective rationale being that they would have learned most of what they were going to learn about manual machines in 10 years. For CNC-specific training, the figures in table 6.5 have been added,

Table 6.6 *Training scores*

Factory	General or career	CNC-specific	Total	Factory	General or career	CNC-specific	Total
J1	0	1	1	B4	10	1	11
J2*	0	3	3	B8	10	5	15
J4	1	3	4	B11	10	2	12
J9	3	3	6	B12*	4	3	7
J45	6	1	7	B39	6	3	9
J50	7	6	13	B71	9	3	12
J66	10	2	12	B145	7	10	17
J140*	0	5	5	B80*	10	5	15
J180	2	10	12	B120	10	8	18

*Indicates large batch factories.

with 'most' being taken as three-quarters, 'some' as one-quarter and 'rarely' as one-tenth. The sums have been divided by the number of operators, and given a number from 1 to 10, with $\frac{1}{1}$ being scored as 1 and $\frac{2.8}{1}$ being scored as 10. The two types of training are given equal weight.

Complete objectivity clearly cannot be claimed. Years of experience on manual machines does not indicate the quality of training, which is obviously difficult to measure, in that period. Furthermore, has OJT been given enough recognition? The measure does not indicate lengths of training periods in the case of CNC–specific training. With regard to underrating OJT, however, recall that the first year of apprenticeships has not been included in the general training figures, hence off-JT may also have been underestimated. Certainly the ages of operators in some cases represents availability rather than preferences, but we are taking a snapshot of the situation and not necessarily measuring preferences. As such, the older operators had had more training and experience on manual machines.

For CNC-specific training, there are arguments both for and against including self-training or self-funded training, since technically it is not training provided by the company. However, it may be argued that employment relations motivate workers to undertake training, hence this type of training should be included, as it has been. The figures exclude family members of J1 and J2, since they represented the owners, and were in a position to be providing the training. Some of the British operators had been recruited externally, but their training backgrounds were not substantially different from internally-trained operators, hence their inclusion does not bias the figures. The scores are given in table 6.6.

The scores do not reflect the whole of the factories' training efforts, of

Table 6.7 *Rank orders: training, OER–MER, size, batch size*

Factory	Training	OER–MER	Size	Batch	Factory	Training	OER–MER	Size	Batch
J1	18	14	18	13	B4	10	18	16	12
J2*	17	14	17	1	B8	3	14	14	8
J4	16	6	15	13	B11	6	12	12	7
J9	14	7	13	16	B12*	13	10	11	2
J45	12	3	9	17	B39	11	8	10	9
J50	5	5	8	18	B71	6	8	6	5
J66	6	4	7	5	B145	2	13	2	9
J140*	15	2	3	4	B80*	3	17	5	3
J180	6	1	1	15	B120	1	10	4	9

*Indicates large batch factories.

course, but training for CNC operators, and while in the British factories CNC operators tended to be those with a lot of working experience before starting on CNC, in the Japanese factories they were younger, with less working experience, and with less training at that stage in their careers.

Training and OER–MER, factory size and batch size

As noted at the end of chapter 4, the weight of evidence regarding the hypotheses rests with the discussions rather than statistical proof. Moreover, representativeness in any statistical sense obviously cannot be claimed. Nonetheless, a non-parametric (Spearman) ranking correlation supports the preceding arguments and will help summarize them. The rankings of the factories in training, OER–MER, factory size and batch size are given in table 6.7.

Table 6.8 gives the coefficients of correlation and significance levels of training–OER–MER, training–factory size and training–batch size for all 18 factories, and for each country separately. As can be seen from the table, there is slight negative correlation between training and OER–MER, but the significance level is very low. In other words, Hypothesis 1 has not been upheld. If anything, more training was given to CNC operators in MER factories than OER factories. There was a degree of correlation between training and OER–MER in the Japanese factories, significant at the .05 level, but the correlation was even higher with factory size (.010). When training was correlated with size for all the factories, the significance level was .005 (i.e., a 1 in 200 chance that the pairings occurred randomly). In spite of very different traditions in training, in both countries the larger factories tended to stress the need for manual machining experience and had the resources to give CNC-specific training, although there were exceptions to this. In Japan, it was the medium-sized factories in conservative indus-

Table 6.8 *Training/OER–MER, size, batch size (Spearman Rank Correlations)*

	OER–MER	Size	Batch
All factories			
coefficient	−.1438	.5937	.0031
N	18	18	18
significance	.297	.005	.495
Japanese factories			
coefficient	.6387	.7531	−.5798
N	9	9	9
significance	.032	.010	.051
British factories			
coefficient	−.2500	.5883	−.2222
N	9	9	9
significance	.258	.048	.283

tries with little growth that stressed experience. In Britain, it was the larger factories which stressed craft training most. Regarding batch size, only in the Japanese factories was there any degree of negative correlation between training and batch size, perhaps because the more conservative view of skill requirements of British managers in the large-batch factories overrode the possibility of having CNC operators with little training and a few skilled supervisors as the reservoir of necessary skills.

The difference between the countries, however, was also significant. The mean training rank of the British factories was 6.6 compared with 12.4 for the Japanese factories. The (one-tail) probability that this was random was .009 according to a Mann–Whitney test. (The same test performed for factory size – 5 largest in both countries versus 4 smallest – and batch size – large batch versus small batch – produces probabilities of .027 and .138 respectively.) Reasons for this have already been suggested during the course of the chapter. They were related mainly to differences in the level of machining skills considered necessary to operate CNC or to deal with unforeseen circumstances. Where more manual machine operating experience was considered necessary, so too was more CNC-specific training. These differences in turn were related to (1) differences in education hence starting levels of operators – which might have meant that operators in the Japanese factories acquired CNC skills more quickly than their British counterparts; (2) different levels of supervision – closer supervision in the Japanese factories meant that less experienced operators could be taught while actually operating; and (3) greater stress placed on adaptability or general intelligence by Japanese managers, and more on machining skills and knowledge by British managers. Of these, the last was probably the most important. While industrial relations considerations played a part – putting semi-skilled or unskilled workers on to CNC would have aroused craft resistance – the view of managers and supervisors, most of whom had done apprenticeships themselves, was that not only did CNC work require these skills, but often that these were best obtained or could only be obtained through apprenticeships. We will return to this view in the final chapter.

7

Division of labour

Much of the controversy surrounding CNC is over who does what tasks, particularly programming. Is, in fact, the 'conceptual' part of operating prised apart from the 'execution' and taken into a programming office firmly under management control? We saw in chapter 6 that skilled operators (in the craft sense) were selected for CNC in the British factories. This does not automatically translate into wider task ranges, however. If indeed Taylorism has been more influential in the British factories, if there is a low trust dynamic and if job-related pay systems and occupation-contoured industrial relations have led to tightly defined job boundaries as argued in chapter 1, operators might have a narrower task range in spite of their experience.

There is more to CNC work than button pushing and programming. Broadly speaking, operator tasks can be divided into three groups; operating, setting and programming. These are tasks essential for machining a part on a CNC machine. There are also a number of related tasks which operators may be asked to do such as inspecting parts after they are machined, and they may also operate more than one machine simultaneously. The first section looks at operating and setting tasks, and related tasks. This discussion is relevant for those interested in what constitutes a 'flexible firm', since we shall also be looking at (functional) flexibility and constraints on flexibility.

The second section focusses on the programming issue. It looks not only at the programming tasks done by operators, but also at the backgrounds of specialist programmers, which will amplify the discussion of mobility in chapter 3, as well as approaches to CNC outlined in chapters 5 and 6. In the third section the factories are evaluated and ranked in terms of task range and the rankings correlated with OER–MER, factory size and batch size rankings, as was done with training in the last chapter.

7.1 Operators' tasks

An attempt to depict operators' tasks in the various factories has been made in figure 7.1. The capital letters A to G represent tasks necessary for the machining of parts, and the small letters represent other tasks which may be done by operators, while parts are being machined, for example. Operators who do only A and B are solely operators, but if they also do C and D, they are operator-setters. If E, F and G are included, they are operator-setter-programmers. The capital letters very roughly indicate the order in which tasks are learned if operators work on manual machines first, but in some of the (Japanese) factories which took operators directly on to CNC, programming was taught to the operators before they knew much about setting up. Setting tasks – selecting the right tools and setting them, setting the parts on selected fixtures – necessitate an understanding of the best way to machine a part, the merits of different tools and fixtures and machining methods, and so on, which those operators did not have. The 'programming first' approach represents a significantly different approach to that of 'machining first' – fast or slow, long or short track – from A and B towards G.

The small letters indicate tasks that are sometimes done by other groups of workers (loading and unloading may actually be put in this category, hence it is placed to the left of 'button pushing', but without it a part cannot be machined, hence it is assigned a capital letter) but may be done by operators. Performance of these tasks represents one kind of flexibility which is receiving increasing attention in British industry but which has not yet been consistently implemented. Figure 7.1 does not indicate the difficulties of the various tasks in the factories, which varied according to batch size, machine type and parts being machined. Setting for large batches of small, round parts on a turning centre at J2 was much simpler than the small batch, stainless parts being machined on a jig borer at J50, for example. In other words, it shows the range of tasks done by the operator, but not necessarily the depth, which illustrates that a discussion of 'skills' must consider more than just task ranges.

Operating and setting

Starting with operating tasks, loading and unloading refers to moving parts on and off the machines, which may involve the use of cranes if the parts are large. Button pushing refers to starting and stopping the machining process, and machine minding to watching the process to make sure that everything is proceeding smoothly. Both A and B were done by all operators, except in

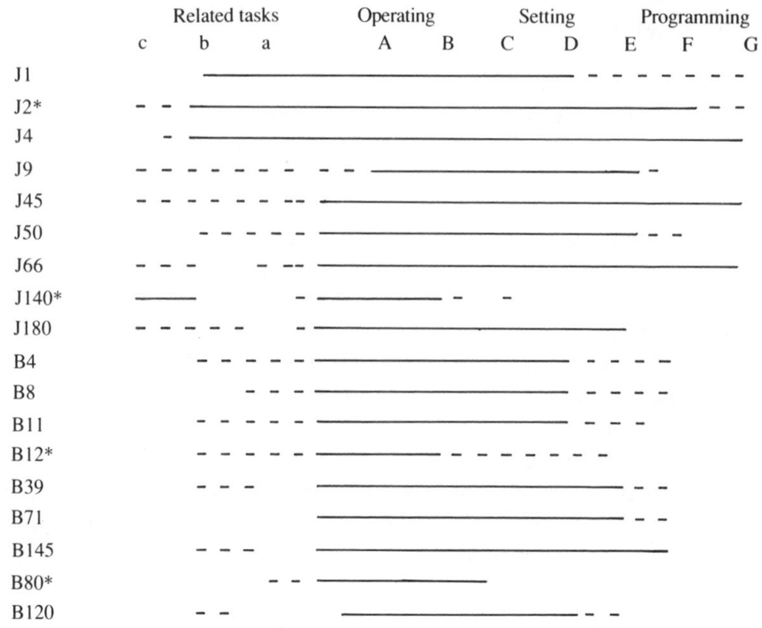

Related tasks Operating Setting Programming
c b a A B C D E F G

J1
J2*
J4
J9
J45
J50
J66
J140*
J180
B4
B8
B11
B12*
B39
B71
B145
B80*
B120

Figure 7.1 CNC operator tasks

Note: a deburring
b inspection
c multi-machine operating
A loading and unloading
B button pushing, machine minding
C tool setting
D machine setting
E programme proving/editing
F simple part programming
G all part programming

Note: Dotted line indicates either task performed on some occasions or performed by some operators.

*Indicates large batch factories.

B120, where there were full time 'slingers' to do loading and unloading. This was not just related to the size of the parts; those at factories like B145 were even bigger. The slingers belonged to a different union (Allied Trades), and this demarcation represented a rigid division of labour at the factory that was only slowly breaking down. At B145 there used to be full time slingers, and craftsmen never did their own loading or unloading. This and other demarcations were abolished after the two protracted strikes in 1978 and

1984. There were also full time crane operators and slingers at J50, apparently because of the size of some parts.

Large batches: operating only

While almost all operators did both loading and unloading tasks as well as simple operating, these formed the bulk of their tasks in J140, B12 and B80; the large batch factories. (The operators of J2 are shown as having an extended task range by virtue of the fact that setting was quite simple; automatic bar loaders fed the lathes, and tooling was also relatively simple.) Conventional machines designed to handle large batches may require special setting skills such as setting cams. Where these tasks were separated from operating, the division persisted on CNC despite the simplification to a certain extent of setting tasks.

At J140 most operators were performing a limited range of tasks, but they were doing them on many machines. The bulk of the setting work was done by group leaders. The limited task range of the B80 operators represents a division of labour between (semi-skilled) operators and (skilled or time-served) setters which was about to be dissolved with the changeover to module production and skilled operators such as had happened in B120. There was no official division in B12, but in practice the time-served operators plus one or two other operators were the only ones trusted to do the setting up. Operators learned to operate one machine, then more machines, and if they showed a lot of promise, they were able to begin to work on setting up.

While there were no out-and-out 'button pushers' in the factories, operators in the large batch factories had a narrower range of operating (capital letter) tasks than those in the other factories.

Small batches: operating and setting

In the small batch factories operators did both operating and setting. In some of the Japanese factories – J1 and J9 for example – operators were just learning how to do setting under the instruction of their group leaders. As explained, tool setting involves being able to decide which tools are necessary for the cutting operations, and machine setting being able to set the workpiece in place with jigs or fixtures (which also implies knowing how they will be cut). Setting ability is normally the hallmark of a skilled worker, depending on the type of machine and parts being machined. However, with more operations able to be carried out with one setting of a workpiece – on machining centres with automatic tool changers (ATC), for example – and with machines being able to cut in several dimensions simultaneously, setting tasks were considered to be becoming easier.[1]

Related tasks and multimachine operating

Related (small letter) tasks included setting up the next workpiece and filling in worksheets, which most operators did, and such tasks as deburring (filing off rough edges), inspecting (the parts machined with measuring instruments), preventative maintenance (changing oil filters, etc.), and even operating other machines. We shall look particularly at the tasks operators did while their machines were machining parts. While more time was available for related tasks where batch sizes were larger or one part took a long time to be machined, there was in fact no clear-cut difference by batch size in the related tasks performed as there was for operating tasks. There were differences between the countries, however.

Britain

British managers were becoming more concerned about utilizing time more fully, and about reducing the numbers of full-time deburrers, inspectors in some cases, sweepers and maintenance staff. While the impetus for this was efficiency and cost saving, it was also seen as a move towards the greater flexibility associated with Japanese competitiveness.

At B80 and at B39 the number of deburrers and sweepers/cleaners had been reduced significantly over the past five years as operators were taking up more of these tasks, while at B120 more self-inspection work was being done. The operators at B12 had been bought some inspection equipment and new brooms (although some of the brooms appeared to have hardly been used), and at B11 and B8 more inspection and some deburring work was being done by operators.

'We've come a full cycle from the 1960s,' remarked the production engineering manager at B71, 'When everything was broken down into little parts.' It would be an exaggeration to say that was happening in practice, however. Some of the operators at B71 did deburring work while their machine was running, others did nothing. The flexibility gained at B145 meant that operators were to do their own loading and unloading, including crane operating, deburring, inspecting, 'and even sweeping the floor and painting if necessary'. According to the convenor, however, this did not happen because 'the company wants skilled workers doing skilled jobs and not wielding paint brushes'. At B4 the MD shrugged: 'Some will go off and set other machines, others will read the newspaper. They look offended if I ask them to do something else.'

In other words, while operators were being encouraged to take on extra tasks if they could, the actual carrying out of these tasks was often left to their discretion. Their first duty was to make sure nothing went wrong with their machine. When it came to sweeping, some operators were more conscientious than others as a matter of pride. As for preventative maintenance, while oiling slides was considered a part of the operator's job, extra tasks were done by maintenance workers. Several supervisors remarked that if it were up to the operators some would do them and some wouldn't.

Japan

In the Japanese factories operators possibly did more inspection work, although there were almost as many inspectors as in the British factories. Deburring was commonly done by the *shiage* group (completion group; this was one job, apart from simple drilling, where one found female workers) when the parts went into assembly. Older, 'retired' workers did the more thorough sweeping up, but sweeping around machines and particularly preventative maintenance tasks were more likely to be seen as part and parcel of the operator's job.

The operator was expected to keep himself busy while a part was being machined; this was company time. Apart from preparing for the next operation, a common task was operating a second machine. Small manual machines were placed near the CNC machine to facilitate second machine operating (e.g. J45), and in some of the factories operators operated more than one CNC machine. This included the large batch factories of J2 and J140, but also factories like J9 and J180. The only factories in which this did not take place were J1, where there were only two operators and two CNC machines in the CNC workshop, and J50, which, as we have seen, had the most cautious approach to CNC work of the Japanese factories.

Multimachine operating

The willingness to carry out parallel or multimachine operating in the Japanese factories and the reluctance in the British factories parallels the discussion of unmanned operating in chapter 5. Occasionally two or even three bar lathes (or planing machines) might be operated in parallel – as at B120 – but it did not happen with CNC machines, nor was that possibility an attraction of CNC machines. In the large-batch factories the notion was dismissed: 'What? And leave it with no-one there? We're getting them to do deburring and inspection work now, but the operator has to be there' (B12). 'We might get away with parallel machining (2 machines), but no more' (B80, referring to union approval). At B145 it was not a matter of unions approving or disapproving: 'I've left open a clause in the agreements to allow for parallel machining, but it hasn't happened yet' (IR manager). Two machining centres had also been put back to back to permit it, but 'company policy' stipulated one man to one machine, according to a supervisor.

The machines were too expensive to be left unattended. An operator could do other tasks during machining time, but he should still be able to keep his eye and ear on his machine.

We have seen that batch size has a bearing on the division of operating and setting tasks. Where batch sizes were large, there tended to be a division between operators and specialist setters or setter-operators. Where batch sizes were small, there were setter-operators. This was more or less true in both countries. The exception was J2, where setting tasks were relatively simple. There were, however, differences between the countries in terms of related tasks. These were more likely to be seen as a part of the operator's job in the Japanese factories, but as an as yet uncertain addition in the British factories. Multimachine operating, too, was more widely practised in the Japanese factories, an observation which parallels the discussion of unmanned operation in chapter 5.

7.2 Programming

Programming includes proving/editing (a trial run for the first part of a new batch, and editing the tape if necessary), simple part programming (writing programmes for simple parts, or filling in the 'flesh' on 'skeleton' programmes), or doing all programming. Braverman argued in 1974:

> The unity of this process (conceptualization and calculation, and metal cutting) in the hands of the skilled machinist is perfectly feasible, and indeed has much to recommend it, since the knowledge of metal-cutting practices which is required for programming is already mastered by the machinist. . . . That this almost never happens is due, of course, to the opportunities the process offers for the destruction of craft and the cheapening of the resulting pieces of labour into which it is broken. (Braverman, 1974, 199)

We have seen that NC/CNC in itself has not brought about a 'destruction of craft' and the replacement of craftsmen by machine minders. Here we will see if in fact operators almost never do programming for the parts they machine. NC/CNC has, of course, been developed technically since Braverman wrote his book, and with the development of manual data input (MDI), background and interactive programming, operator programming has become a more attractive option for some managers.[2] Which ones, will be discussed shortly.

Who does the programming?

Where operators did only proving and editing (E in figure 7.1), or no programming at all, there were specialist programmers who made the programmes in an office. Some were full-time programmers and some did it part time, some were located on the shop-floor and some off it. Where operators did simple part programming, or modifications of old programmes, specialist programmers normally did the more difficult programmes or new programmes. Sometimes, however, operators did all the programming.

Again, batch size had some influence. Of all the large-batch factories, it was only in J2 that operators did any programming at all. The operating and setting tasks at the other large-batch factories were by and large separated, and if any shop-floor workers were involved, it was the setters.

In the other factories the organization of programming tasks was related to personal views of production engineering and production managers. One might cite small batches as a reason why operators should do their own programming (B145) and others as a reason they shouldn't (B39, B11). Those who supported operator programming stressed that the operator was nearest the machining process and knew the machine best, while those who supported specialist programming emphasized the distractions on the shop-floor and increased downtime if operators did their own programming. These managers were influential in the organization of programming tasks. The outcome of these opinions was most visible in the larger factories, since these had separate production engineering departments where programming could be done. In the smaller factories too, however, programming could be done by a foreman or specialist in the foreman's office, as well as by operators.

There was a tendency to greater operator programming in the Japanese factories and more specialist programming in the British factories, but the differences were by no means clear-cut, particularly since there was considerable *de facto* programming by British operators even where there were specialist programmers. It is also useful to divide the discussion into factory size and country.

Japan

The owners' sons in the *smallest* Japanese factories, J1 and J2, had taught or were teaching the other operators to programme. These operators did not need to know how to operate manual machines, but programming was a part of their jobs which they should master at the same time or even before learning setting tasks. At J4 also, it was a natural part of the operator's job.

The foreman had always been too busy to go on a course and no-one else in the company knew anything about programming:

> If they come to me with questions about programming, I say you should know, you went on the course. I try to teach them about speeds and feeds and materials. They pick up, X, Y and Z themselves pretty quickly.

In the *other* factories there were a variety of situations. At J9 the managing director thought that although operators were perfectly capable of doing programming, and did do a little bit at times, it was more efficient to have programmers. The operators did other jobs while their machines were running, anyway. Proving and editing were done by the programmer and operator sometimes, or by the leading hand and the operator.

At J45 there was one programmer, but most of the programming was done by the operators. The production engineering manager claimed they had had that policy from the beginning. 'The staff don't know the machines very well. Yes, the uptime drops a bit if the operator does the programming, but the operator would have to check the programme, anyway.' At J50, which we have seen to be most cautious over CNC use, the production engineers argued that it was too much to ask the operators to do complicated calculations as well as operate. Adaptions of previous tapes were often done by the operators, however, and the operator on the newest machining centre claimed he did most of the programming himself. At J66 the programming was originally done by the production engineers, but it was handed over to the operators in 1980. The production planning manager thought they were capable of doing it, and that it would help to raise morale. By then they had quite a library of tapes to draw upon, which needed only minor modifications for new batches. Finally, at J140 and J180 programming was done by specialist programmers. At J140 the operators had no input into programming at all. They were too busy watching their many machines for that. At J180 there were specialist programmers to maximize machine uptime. 'A lot of the work had difficult contours which would be difficult for the operator to programme while operating. Some of the proving and editing, though, is done on the shop floor.'

In the smallest Japanese factories, then, programming was seen as a natural part of the operator's job, and in fact programming knowledge was more integral than knowledge of conventional machining. In the other factories there was a variety of situations. In J45 the bulk of the programming was done on the shop-floor, while at J66, all of it was. At cautious J50, however, and at expansion and efficiency-conscious J9 and J180, most was done off the floor, while at J140 all was done off the floor.

Division of labour

Britain

In the *smaller* British factories, programming was mostly done in the foreman's office, although some programming was done by the operators. At B4 the man who championed CNC as foreman and who was recently promoted did most of the programming in his office, where he also did estimating and planning. Simple parts, though, were programmed by the operators. At B8 most of the programming was done in the foreman's office by the foreman or the chargehand. The foreman said he was trying to encourage the operators to do simpler programming beyond proving and editing, but two older operators were struggling. At B11 the programming was done in the foreman's office by a computer-wise ex-operator who was being groomed for the foreman's job. A milling/drilling machine, however, was programmed by the operator, and the other operators did their own proving and editing when the programmer was busy.

In the *other* British factories, from B12 up (except for B39) there was a production engineering department, and programming was done in these. At B12 there were four specialist programmers, based in production engineering. As at B11, the shop-floor was said to be too noisy for programming, and operator programming would reduce uptime, i.e., make the (expensive) machine idle for longer periods of time. Proving and editing of tapes, however, was done by the setters on the shop-floor. In B39 it was also said that operator programming would reduce uptime, but the two programmers were very busy and 30–40% of the programming (for simpler parts, or programmes made from 'skeleton programmes') was done on the shop-floor.

The production engineering manager at B71 had a distrust of operator programming; 'MDI? Over my dead body, so to speak... Actually, I have a distrust of operators – on behalf of the company – being able to control their own speeds and feeds. They're on individual incentives, remember. My programmers are planning engineers and tooling engineers, and the programmes are prepared in a satellite environment in here.' The operators had a different view of who did most of the programming, and claimed they virtually had to rewrite some programmes. Quipped one, 'Oh, they rough it out up there.'

On the other hand, the managers at B145, the first factory to purchase NC, were forthright about the merits of operator programming, including the production engineering manager: 'With small batches, there is quite a lot of programming to do. A competent operator with MDI can produce programmes more quickly than we can up here with the complexity of our work. The time the operator spends programming includes finding special tools, proving and editing. Even if we make the programmes up here, those things have to be done.'[3] In 1982 there were 13 programmers, in 1987 there were seven, and this trend was expected to continue. One superintendent was delighted; 'When NC came in our operators lost their pride. Now they can be craftsmen again.'

At B80 the setters proved the tapes with the programmers or by themselves, but the operators were not involved at all. It was said that the AEU members once voted against getting involved in programming, a decision which they later regretted. At B120 the bulk of the programming was done by 10 programmers in manufacturing engineering, but 13 methods engineers also spent about 25–30% of their time programming. The operators were not supposed to do programming. This was another demarcation. There had been some TASS/AEU friction in the past over the matter, and what the programmers wanted was that the operators assist in proving but no more. Some operators in fact did do more.[4]

Some of the managers in the British factories did not seem to know about background programming. Most thought it was best to have the operator involved in operating-related tasks or deburring or inspection while parts were being machined, and also in proving and editing, but not full programming. The shop-floor was too noisy and distracting (from machines and other operators), hence it was best to do most of the programming away from the shop floor, which meant in the foreman's office in the smaller factories and in the production engineering office in the larger factories. In practice, though, the operators were involved in a certain amount of programming beyond proving and editing. Sometimes this was encouraged, in other cases it was tacitly accepted.

Programmers' backgrounds

The backgrounds of specialist programmers also tell a significant story with regard to the broader division of labour in the factories and attitudes towards CNC. Table 7.1 shows that in only one Japanese factory (J180) were there any programmers who were originally based on the shop-floor, and even there the practice of promoting operators had been discontinued. Just as operators were being taken directly onto CNC because of work pressures, the same applied to programmers.

Only one other Japanese programmer (at J9) had had experience on the shop-floor. He was originally assigned to production engineering, but was put on the shop-floor for two years before going back to the office to do programming. The others had never worked on the shop-floor. Some were university graduates and some high school graduates, and a recent tendency in some factories (J180, J140 and J45) was to use young female workers with no engineering experience to do the programming.[5]

Table 7.1 *Programmers*

Factory	No. of CNC	No. of programmers	No. from on shop-floor	Average years on shop-floor	Factory	No. of CNC	No. of programmers	No. from on shop-floor	Average years on shop-floor
J1	2	0	—	—	B4	3	½	½	20
J2*	6	0	—	—	B8	4	½	½	14
J4	7	0	—	—	B11	3	1	1	4
J9	28	4	0	0	B12*	17	4	4	6
J45	15	1½	0	0	B39	6	2	2	8
J50	3	1	0	0	B71	18	2	2	5
J66	10	0	—	—	B145	39	7	7	5
J140*	181	6	0	0	B80*	16	5	3	15
J180	100	20	13	5	B120	30	23[a]	13	15

Note: (1) No. of CNC – number of NC/CNC machine tools.
(2) No. of programmers – number of specialist programmers.
(3) No. from shop-floor – number of programmers originally from shop-floor.
(4) Average years on shop floor
*Indicates large batch factories.
[a]13 spend only 25–30% of their time programming.

While there were few programmers in the Japanese factories with a shop-floor background, there were very few in the British factories without one. Those who hadn't in B120 were involved in direct numerical control (DNC) work, and the two in B80 had come to programming via a technician – draughtsman – tool draughting – methods – programming route. Some of the programmers at B145 had originally been fitters, but all the rest had been operators. It was thought that programmers had to have sound experience and knowledge of operating, machining methods and materials in order to make programmes, whereas in the Japanese factories it was thought that they could learn what they needed to know from other programmers or books, or sometimes from (programming) courses at the manufacturers. There was a further difference in that while specialist programming was a job for young production engineering recruits in Japan – usually the first job they were assigned to – most of the British programmers had done other production engineering jobs beforehand such as work and methods study, and sometimes jobs like production control, in addition to those they had done on the shop-floor. In those Japanese factories where traditional operating skills were most valued, the operators on the shop-floor tended to do the programming also, although at J50 it was done by a graduate engineer (and not by someone without an engineering background). Attitudes towards programmer requirements in the two countries thus paralleled what was found for operators.

Operator–programmer friction

As noted in chapter 3, there were relatively few ex-shop-floor workers in the production engineering offices in Japan. Despite the fact that the programmers in Japan had different backgrounds from the operators, there was very little friction between them of the type that is often reported in British factories. The potential for friction was well recognized, and measures to avoid it ranged from assigning operators and programmers to the same room of the company dormitory (this had happened at J50) to putting them under the same foreman and having them do recreation together. It was sometimes suggested (e.g. at J9) that if operators and programmers had the *same* backgrounds, it might increase the number of arguments rather than diminish them.

While a lot of the complaints about programmers by operators and vice versa in the British factories were not particularly hostile, the fact that they originally had the same background did little to remove the 'Us and Them' divide that was frequently mentioned: 'Most of the ones who've gone over were useless on machines. Sometimes I think that's all they wanted to do – go over into a white collar job. It's definitely an Us and Them situation.' At

B145 rancour between the two groups had subsided. The programmers had gained more experience, while operators on the floor were doing more programming and said they appreciated the problems of those up in the office. The point made, however, is that the lack of an 'Us–Them' divide in the Japanese factories was not the result of job mobility. One therefore has to look to aspects of employment relations – common orientation programme and recreation, common payment and promotion systems, communications and 'company consciousness' – and possibly to a propensity to be less openly critical of other people.

7.3 Task ranges, OER and MER

As training was scored, ranked and correlated with OER–MER rankings in chapter 6, the same will be done here with the task ranges of CNC operators. The prediction of Hypothesis 2 was:

A wider range of tasks is performed by CNC operators where employment relations are organization-oriented than where they are market-oriented.

We shall also consider the third hypothesis:

The influence of factory size and batch size on CNC use is mediated by employment relations.

Task range scores

The same categories are used for scoring task range as were used at the beginning of the chapter, namely, operating tasks, setting tasks, programming tasks and additional or related tasks. A total of 10 points will be given for each of these categories. The categories are further subdivided, as in figure 7.1; and where there are two tasks in a category, each is given a maximum of 5 points, where there are three, the task most closely related to operating tasks is given 4 points and the others three. Again, a snap-shot view has been taken in which some operators were in the process of learning new tasks. Lower scores were given in these cases, since in effect other people were doing part of those tasks. The evaluations of the related tasks are slightly more subjective than the others because of difficulties in classifying answers and limited observation time, but by and large the overall scores match reasonably well with responses given in interviews and with observation. The scores are given in table 7.2. An obvious objection to scoring factories in this way is that operating, setting, programming, and even related tasks were not comparable in the factories. Setting tasks at J2, for example, were simpler than in many of the other factories, and the same

Table 7.2 *CNC operator task range*

	Related Tasks			Operating Setting				Programming			Total	Rank
	c	b	a	A	B	C	D	E	F	G		
Maximum	4	3	3	5	5	5	5	4	3	3	40	
J1	0	2	3	4	5	4	4	3	3	2	30	5
J2*	3	2	2	5	5	4	4	3	3	2	33	1
J4	1	2	2	5	5	4	4	4	3	3	33	1
J9	2	1	2	5	5	4	4	4	1	0	28	6
J45	3	1	0	5	5	5	5	4	3	0	31	4
J50	0	1	1	3	5	5	5	4	1	0	25	12
J66	1	0	1	5	5	5	5	4	3	3	32	3
J140*	4	0	0	5	5	1	1	0	0	0	17	16
J180	2	1	0	5	5	5	4	4	0	0	26	10
B4	0	1	2	5	5	5	5	3	1	0	27	8
B8	0	0	1	5	5	5	5	3	2	0	26	10
B11	0	1	1	5	5	5	5	2	1	0	25	12
B12*	0	1	1	5	5	2	2	2	0	0	18	15
B39	0	1	0	5	5	5	5	4	2	0	27	8
B71	0	0	0	5	5	5	5	4	1	0	25	12
B145	0	1	0	5	5	5	5	4	3	0	28	6
B80*	1	0	1	5	5	0	0	0	0	0	12	18
B120	0	1	0	1	5	5	5	2	0	0	15	17

Notes: A – loading and unloading
B – button pushing, machine minding
C – tool setting
D – machine setting
E – programme proving/editing
F – simple part programming
G – all part programming
a – deburring
b – inspection
c – multi-machine operating
*Indicates large batch factories

could be said of J1. The score at J50 might have been higher if operators did not have so much stainless work, or if the pieces were smaller. That product type did not determine task range, however, can be seen from the different scores of the 'paired' factories: J180 and B120, J140 and B80, J66/J50 and B145/B71, and J45 and B39, for instance. It can also be seen by the fact that the parts J4 produced were not particularly simple compared with the other Japanese factories, yet it shares the highest score with J2. Machine type obviously had some influence, but this was not decisive.

The variable in question is task range and not skill *per se*. If task range were the sole criterion for judging levels of skill – which is how the issue is often framed in the deskilling debate – we might conclude that the operators of J2 and J4 were the most skilled, but their lack of experience on both manual machines and other types of CNC machines, and their limited setting skills (for other types of products, for example), would prevent them from being able to do many of the tasks that a J50 or B39 operator would be able to do. This is why any overall evaluation of skills must take into consideration the discussions of both chapter 6 and chapter 7. If we look just at the range of tasks, we indeed find that increasing factory size and batch size are related to narrower task ranges as Sorge *et al.* (1983) found, but that in itself does not give a full picture of the deployment of workers for CNC work. We can see from the scores that those factories in which operators did the broadest range of tasks were not necessarily those in which the most training was given.

Task range and OER–MER, factory size, batch size and training

Spearman's rank correlation test is again used to judge the degree of correlation between task range and OER–MER, task range and factory size, task range and batch size, and task range and training. The coefficients of correlation and significance levels are given in table 7.3.

Again, and contrary to Hypothesis 2, there is no significant correlation between the OER–MER rankings and those of task range. There was a significant (negative) correlation between factory size and task range in the combined factories and the Japanese factories, though not for the British factories alone, where the negative correlation with batch size was much more significant.

The Sorge *et al.* thesis of factory size and batch size shaping the organization of tasks seems to be a more powerful predictor than employment relations. That thesis, it will be recalled, was that increasing factory size is related to a separation of functions such as planning and programming, while increasing batch size is related to differentiation of operating from

Table 7.3 *Task range, OER–MER, size, batch size (Spearman Rank Correlations)*

	OER–MER	Size	Batch	Training
All factories				
coefficient	.0722	−.5067	−.2885	−.5034
N	18	18	18	18
significance	.388	.016	.123	.017
Japanese factories				
coefficient	−.5000	−.5941	−.3487	−.4118
N	9	9	9	9
significance	.085	.046	.179	.135
British factories				
coefficient	−.0763	−.1933	−.6581	−.1441
N	9	9	9	9
significance	.423	.309	.027	.356

setting and programming. Some reservations must be expressed, however. First there is no significant correlation of factory size and batch size combined with range of tasks.[6] Secondly, the fact that there is a significant negative correlation between task range and batch size in the British factories, but between task range and factory size in the Japanese factories, is of some importance.

The British operators performed a narrower range of tasks (the significance of the country differences is .006 according to a one-tail Mann–Whitney test, while the corresponding figures for factory size and batch size groups are .091 and .098 respectively), even though they scored very highly on training. The biggest differences between the countries in task ranges derived from related tasks; British operators performed less of them. A number of reasons might be suggested for this. First, Taylorite, or more precisely, Babbage[7] principles of decomposing skilled work were in greater evidence. Craftsmen retained the most skilled parts of their jobs, and the rest were performed by semi-skilled or unskilled workmen. Skill or job-related wage determination reinforced the separation. The outcome is most clearly seen in large-batch work, where not only related tasks were separated from operating–setting, but operating and setting were separated also.

In Japan, on the other hand, the development of employment relations has led to less significance being placed on the task range of a given job (whether in terms of pay or control). Operators may be required to perform a wide range of tasks even where batch sizes are large, and it is factory size, not batch size, which produces the most significant (negative) correlation. The task range in small factories, however, is greatest not necessarily because of less bureaucratization of jobs, as Sorge *et al.* suggested – after all, foremen could have done the programming there as they did in the smaller British factories - but because of the way the managers in these factories viewed CNC (which we will return to in the next chapter). Thus variables such as size and factory size do influence task ranges, but this influence is modified by aspects of employment relations in both countries.

Intrajob flexibility and interjob mobility

British operators did have a narrower task range than their counterparts in the Japanese factories, where the concept of a 'job' was more flexible. This should not be confused, however, with movement between jobs. British CNC operators tended to be the 'cream' and on their way up they experienced a number of jobs. It could be anything up to ten years before an operator changed jobs even at J180 in Japan, where rotation was planned. We also saw how a manager in J45 told his unhappy young laser machine operator that he would shift him in *two or three years* if he were still unhappy then.

While British operators did not operate more than one CNC machine at the same time, if another worker was absent, they were sometimes moved if there was priority work elsewhere. The operators usually welcomed the chance to go back onto manual machines for short periods so that they would not forget their manual skills, although some said they would 'die' if they had to go back onto them permanently. Thus there was flexibility in short-term movement, and pay disparities had been worked out so that the operator received whichever rate was higher where there were different gradings. It should not be assumed, therefore, that less intrajob flexibility automatically means less interjob movement or vice versa, even if there are institutional obstacles to interjob movement to be overcome.

8

CNC use and skills

8.1 Craft versus technical approach

One of the attractions of CNC for Japanese purchasers was their potential for running unmanned, or to have one operator operate more than one machine simultaneously. For British purchasers, the most responsible approach was to have a skilled operator by the (single) machine, which very often meant having a second shift. In selecting operators Japanese managers placed a strong emphasis on general brightness, maths ability and adaptability, whereas most of the British managers stressed machining skills and knowledge for both operators and programmers. These and other related differences suggest a qualitatively different approach to CNC in both countries. The major differences, somewhat exaggerated, are listed in table 8.1.

The observations in the first column suggest what may be called a *technical approach* to CNC, and those in the second column a *craft approach*. In fact, in some of the Japanese factories, slogans emphasized a shift from craft-type skills (*gino*) to technical skills (*gijutsu*); 'From hand work to head work'; 'From KKD (inspiration, experience[1] and compromise) to NTT (numerical, time and technique),' from an old 'Orthodox' to a new 'Orthodox' and so on. One aspect of QC circles, too, was the greater application of 'scientific' principles to production work. Even if the 'intellectualization of blue collar work' was reflected more in slogans than in actual work organization, the consciousness of such a shift and the principles involved had permeated the factories to shop-floor workers.[2]

There were, of course, differences amongst the Japanese factories, which may be linked to labour market, training, and to some extent, product differences. Related to the technical and craft approaches, but not identical, are what can be termed a *programming first* and a *machining first* approach. Expressed in an extreme form, this is the difference between treating a CNC machine as a computer with a machine tool attached and treating it as a

Table 8.1 *Differences in CNC use and attitudes to CNC: Japan and Britain*

Japan	Britain
Chapter 5	
Unmanned operating seen as an attraction of CNC, and unmanned operating carried out.	Unmanned operating not seen as an attraction of CNC; an operator had to be present.
Chapter 6	
Operators often had little or no experience on manual machines.	Operators had considerable experience on manual machines, and often craft backgrounds.
Operators chosen for CNC given relatively less CNC-specific training, and were sometimes required to master it themselves.	Operators chosen for CNC given relatively more CNC-specific training.
Chapter 7	
Multimachine operating seen as an attraction of CNC; less experienced operators could operate more than one machine.	Multimachine operating not seen as an attraction of CNC; an experienced operator had to concentrate on one machine.
Specialist programmers not required to have machining experience; almost none had a shop-floor background.	Specialist programmers required to have considerable machining experience; almost all had shop-floor backgrounds.

machine tool with a computer attached. It was the former approach that got an installation engineer evicted from B11, and a number of other British managers complained of the 'programming first' approach of training courses at skill centres and even manufacturers. Some of the managers in the medium-sized Japanese factories were dubious about this approach, too.

It was strongest in the smallest factories, where there were difficulties in attracting and keeping young workers, limited training resources and pressures to keep down labour costs. Young workers were taken directly onto CNC, were taught programming at an early stage, and were not expected to learn how to operate manual machines. They were to find information about speeds and feeds in handbooks. If they left, another could be trained in a matter of weeks because they had learnt about computers, anyway. *Machi koba* ('street corner' factory) lathe operator and author Koseki talks about the 'CNC language speakers' (*NC gengozku*) and the 'machine language speakers' (*kikai gengozoku*).[3] The former represent a generation being raised who do not know the language of machines, and there is a shortage of interpreters for the two groups. Koseki's MD commented once:

> Look at the primary school children these days. They're playing around with personal computers. If you don't give young workers something like that to do, if you keep them on filing, they'll quit. You've got to let them learn it like a game.

The larger factories did not have the same problems in attracting and keeping young workers, were more concerned with long-term training and less with reducing labour costs. To varying degrees the operators were expected to learn manual machining skills. At J140 operators were taken onto CNC after half a day's instruction. These workers, however, were expected to acquire manual machining skills for the skill tests, which were semi-obligatory. While managers at the larger factories did not adopt a 'programming first approach' and thought that manual machining skills were necessary, even in the most conservative factories (in terms of views of CNC skill requirements) at least the slogans for a shift from craft skills to technical skills were evident. For these managers, too, a CNC machine tool was basically an obedient, technical machine that would only misfunction as the result of human error, not a whimsical machine ready to spring a surprise on the unwary.

In the British factories the 'machining' first approach dominated, although there were differences in the degree to which machining skills were equated with a craft background. While a shift to technical skills is sometimes talked about in Britain, there was little evidence of it or consciousness of it in the factories. The 'instinct' and 'experience' that were to be replaced in some Japanese factories were the very attributes that many British managers valued. The operator who could, as a result of his long experience on machines or craft background 'see problems on the horizon', 'feel when something is going to go wrong' or 'listen to the sound of the machine and know if it is cutting all right' was the right person for the job.[4]

8.2 CNC, skills and deskilling

Hypothesis 4, it will be recalled, predicted:
Skill levels are preserved or enhanced where employment relations are organization-oriented, and contested where they are market-oriented.

Can either approach be said to incorporate a dynamic of 'deskilling'?

'Deskilling' can refer either to jobs or people. Referring to jobs, a number of surveys carried out in Japan show that managers think it takes considerably less time for a new operator to become competent on a CNC machine than a conventional machine; one year as opposed to four years according to one NIEVR survey (1983, 27); 10 years has now become 5 years according to a MITI study (Tsusansho, 1984, 18). Most observers agree that CNC has simplified tooling and setting tasks, wherein much of the conventional skill lies. Predictably, the MITI study suggested that the work had become easier manually, but more difficult mentally (Tsusansho, 1984, 72–3), a view which many of the operators in this study – both British and Japanese – concurred with, usually approvingly.

Whether or not jobs are 'deskilled' (or 'reskilled') as a variable within factories, however, depends upon the tasks operators are expected to do. If they are expected to do operating tasks only, for example, where they once did setting tasks as well, the jobs may well have been 'deskilled', and if programming had been separated from operating and setting, there is also a case for 'deskilling'. The intentions of the organizers of tasks must be examined, however, before one can talk about 'deskilling strategies' (or obsessions). These should not be assumed.

Separation of operating and other tasks was related, as hitherto, to batch sizes, and where setting tasks were already separated, so too were programming tasks. There was somewhat more operator programming in the Japanese factories, but the choice to have specialist programmers in both countries was prompted particularly by considerations of machine uptime. As Francis (1986, 77) points out, profits are realized not only by extracting increased surplus value from labourers. Turnaround time (time from order to delivery), hence machine uptime, is an important aspect not only of production costs, but also for gaining new orders. Reduction in turnaround time was one of the attractions of CNC in both countries.

There was little evidence that specialist programmers were being used to wrest control of the work process away from craftsmen. Of the British factories, even at B71, where the production engineering manager expressed a distrust of operators doing their own programming, the operators did in fact do a considerable amount, which was recognized by a £2 per week premium. At B120, where operators were not supposed to be involved in programming beyond proving and editing (there had been TASS/AEU friction), some operators did do their own programming. At B80 the operators were about to be *upgraded* rather than the reverse, and at B145 the programming department was shrinking, not because of pressure from the traditionally militant unions, but because managers thought that operators, who were closest to the work process, made the best programmers.

Regarding people, one might argue along with Koseki that both skill requirements and operator skills have constantly undergone change, with changing materials, machines and tools, hence the concept of a 'skilled' worker needs to be viewed dynamically rather than statically: 'A skilled worker these days is completely different from when I started out [1950]. Making your own tools was an important part of the job then, but it's not nowadays. Skills change; a workman's skills are dynamic, not static.'[5]

Whether or not *operators* are 'deskilled' (or 'reskilled') as a variable within factories depends on who is chosen to operate CNC machines. We saw that in the British factories very skilled operators were chosen, although some of the larger factories were tempted to use semi-skilled operators at first. There was no subsequent attempt by managers to supplant skilled operators with semi-skilled or unskilled operators, although some predicted that skilled operators would eventually become bored with CNC and be happy to relinquish operating tasks to less skilled workers. Whether skilled workers lost their skills after being put on CNC is another question. Most were happy that they could work on manual machines again occasionally, but a common sentiment was that they preferred CNC work, as expressed by the senior AEU steward at B71:

> I never thought about it [being deskilled] until B [industrial relations manager] asked me the other week if I thought I'd thrown my skills away with CNC. I'd rather work on CNC. It's the thing of the future.

What about the Japanese factories which chose inexperienced operators for CNC, often passing by more experienced operators in the process? The managers in the smallest Japanese factories were very conscious about saving labour costs, and their strategy of using young workers whose wages would be low might be considered a 'deskilling' strategy, although for economic reasons, rather than control or class objectives. In part, however, this strategy derived from the widespread belief that young workers were more adaptable, and more suited to the new machines.

The shift from a craft approach to a technical approach was fuelled by the belief that this was a historic trend, and a key factor in competitiveness, and a failure to adopt this approach would result in being left behind. To describe this as a 'deskilling strategy' at the larger factories would not seem appropriate, especially in view of the fact that not only was control over the work process not an issue, but having younger, less experienced operators would hardly save on labour costs given that it was difficult to lay off older workers (although they might eventually be 'loaned' out).

8.3 Employment relations and new technology

The technical and craft approaches are not necessarily the result of conscious and consistent policies. It would be misleading, therefore, to describe them as alternative 'strategies'. They also reflect a different degree of management involvement in a traditionally craft-oriented work process. Japanese observers of British industry often comment on the less direct involvement of management in both employment relations and the work process. Totsuka *et al.* (1987) use the terms *nariyuki kanri* (laissez-faire management) and *kanri fuzai* (management 'not at home').[6] By comparison Japanese managers are more directly involved in shaping both, although they will delegate responsibility where possible within the parameters they define. Greater involvement in employment relations can be seen for example, in orientation training and the evaluated component of pay. Greater involvement in the work process can be seen in attempts to bring 'experience' and 'inspiration' into the realm of scientific analysis and problem solving. It is quite likely that greater involvement in employment relations facilitates greater involvement in the work process.

It might seem that Bravermanian deskilling as the result of management intervention and replacement of craft skills by 'scientific, technical and engineering knowledge' is characteristic of Japanese factories. The objective, however, is primarily one of increasing predictability and thence efficiency rather than deskilling workers. Furthermore, workers are not being excluded from this 'scientific, technical and engineering' knowledge but are being urged to acquire it.

They are supposed to do this through self-study, QC circles, and also through interaction with other groups of workers. Hattori (1986, 321) claims that there is a smoother flow of 'scientific, theoretical and universal' knowledge from production engineering departments to the shop-floor and 'concrete, experiential and individual' skills from the shop-floor to production engineering, often via the production control department, in Japanese firms than those in other countries. He argues that there are three main facilitating conditions for this virtuous spiral of technology transfer within (and between) organizations: good communication, job security and a sense of common destiny, and a lack of clearly defined job boundaries.

Facilitating conditions are not always sufficient conditions; the skills of the operators in this study were not necessarily the 'advanced technical' or 'production engineering type' skills that are sometimes claimed for them.[7] Nonetheless, there did seem to be such an information flow despite the lack of personnel interchange. The comments of a production engineer at J50 captures this: 'We actually have quite good relations. We go out there and talk things over with them, although we don't tell them how to run their machines. Our office is not on the shop-floor now, but we don't feel distant. Also, we have recreation and sports activities with them.' Despite the fact that many production engineering workers were originally from the shop-floor in the British factories, there may well have been less of an information flow, at least in the reverse direction. Shop-floor workers mentioned an 'Us–Them' divide (and shop-floor workers who crossed it belonged to 'Them'). There were instances of production engineers training operators in CNC use (for example at B120), but it is doubtful that they were trying to impart an engineering or technical approach to CNC, or if this would have been welcomed.[8]

While 'harmonization' might weaken the divide, it is doubtful that it will eliminate it and replace it with cooperation. One reason is that the harmonization measures themselves do not go as far as earlier reforms in the (larger) Japanese factories towards employing different groups of workers for membership in the same organization and blurring the divisions between those groups in employment, payment and industrial relations as well as with a 'common community of fate' ideology. This is another illustration of the interrelationship between employment relations and the work process (including the use of new technology), and the degree of management involvement in these.

A third aspect has already been mentioned. It was suggested in chapter 5 that slogans heralding a shift in skill bases and calling for renewal in the larger Japanese factories were not just an accompaniment to technological change, but an attempt to galvanize the attention and energy of the employees towards common goals and the future. Failure to innovate would not only jeopardize company performance, but would also result in flagging morale, and the two would reinforce each other in a vicious circle. It was also suggested that similar attempts by managers in British counterparts would be greeted by passivity if not resistance because of their employment relations.

Employment relations therefore do influence approaches to new technology, but the influence is more subtle than that suggested by the hypotheses. The craft and technical approaches are shaped by employment relations and degree of management involvement, but all of these should be seen in their broader historical and social contexts. For example, the organization of craft workers and craft training have themselves influenced the development of employment relations and degree of management involvement in Britain, which have in turn influenced craft training and organization. These have reinforced a certain view of skill requirements both among workers and managers. In Japan, on the other hand, there has been a different dynamic. The limited craft organization and training was no doubt one factor in both employment relations and training developing along a different path than in Britain, and the interrelationship of these has led to a different view of skill requirements.

The wider society in which initial education and socialization takes place is also influential. Mention was made of general education, in which it seems that a higher level of 'three R' (particularly mathematics) competence is imparted in Japan to those who will become operators, and this facilitates the technical approach. The present government in Britain has undertaken reforms in education, one objective being to raise mathematics standards, but it is not clear what the outcome will be. Given other priorities such as introducing greater (market) choice, one might predict that here, too, the 'cream' will continue to rise, and a major objective is to remove impediments to that process. If so, major changes in the backgrounds of operators cannot be expected.

The aims of this book have been descriptive rather than prescriptive, but the question naturally arises today as to which of these approaches to CNC is more effective in a competitive sense. Given the strong competitive ability of many Japanese firms, one might answer in favour of the technical approach, citing, for example, unmanned operating in lieu of shiftwork, which was seen by many Japanese managers as expensive both in a monetary and a social sense. The technical approach and an emphasis on technical skills might also facilitate the adoption of more advanced forms of automation. On the other hand, Sorge and Warner (1986, 163) note that an emphasis on craft skills with CNC use is linked to economic success in West German firms, also with strong competitive ability.

Given historical, social and educational backgrounds, as well as employment relations and the level of management involvement, wholesale adoption of the technical approach in British factories would be neither advised nor feasible. That is not to say, however, that it should be ignored. At the very least, managers – and unions – should be aware of both approaches. They should also be aware of the characteristics of the different employment relations systems which have been described. Selective learning from other countries which builds on existing strengths can enhance competitiveness, as the Japanese experience has shown.

Finally, if the technical approach does offer the potential for competitive advantage, it is not an automatic consequence. Idle CNC machines can be found in Japanese factories as in British ones, as when there is a sick operator and no-one to replace him. One occasionally comes across 'window display' CNC machines which were never mastered, or bought for 'public relations' purposes, too. Again logical possibilities do not always lead to expected outcomes, and care must be taken to separate the two. A failure to do this has resulted in considerable confusion both in research on new technology and in cross-national research.

Appendix 1 Glossary of technical terms[1]

Boring: machining process whereby a rotating tool cuts an internal surface to dimensions normally larger than the tool itself.

CAD/CAM stands for computer aided design and computer aided manufacture. An integrated system of producing components from the design stage through to manufacture using computer assistance.

CNC stands for computer numerical control. An NC (see below) technique which utilises a dedicated stored-programme computer, hence makes it possible to store programmes, edit those programmes, etc.

DNC: the downloading of a part programme from a host computer directly into the memory unit of a CNC machine tool or tools.

Drilling: machining process whereby a rotating bit is introduced into a workpiece to cut a hole into or through it.

FMS stands for flexible manufacturing system. A means of organizing production utilizing at least two CNC machines, with automatic tool and component changing according to a (modifiable) programme.

MDI stands for manual data input. The keying-in by hand of a part programme into a CNC control unit via the unit's console keyboard.

Milling: machining process whereby a fixed workpiece is introduced to a rotating tool to produce grooves, bevels, flat or contoured surfaces.

NC stands for numerical control. Control of machining functions by means of a predetermined code consisting of numbers, letters and other symbols and initiated via an electronic control system.

Planing, shaping: machining processes whereby a fixed workpiece is introduced to a stationary tool and back again to produce a cut (planing), or a tool with a reciprocating movement is fed along a surface and back again to produce a cut (shaping).

Turning: machining process whereby a tool held rigidly in a tool post is introduced along the axis of the lathe bar to a rotating workpiece.

Appendix 2 The 18 factories: a brief introduction

Below is a brief introduction to the 18 factories. Some of the information on the factories came from books, magazines, company newspapers and reports, and union reports (especially in Japan). Most, however, came from interviews and observations conducted during 1986–7 with some follow-up visits in 1988. At least two days were spent at each factory, where personnel managers or senior general managers (where there was no personnel manager), union representatives, production engineering managers, programmers, production managers, foremen and operators were interviewed. In some cases one person filled more than one category, for example, union representative and programmer.

The importance of shop-floor observation and interviews was clear from another small Japanese factory excluded from the final list, where the MD spoke at great length of their six CNC machines, the training given to workers for CNC, and so on, while a puzzled foreman later figured out that he must have been talking about their 'programme control' (somewhat akin to point-to-point NC) jig borers with digital control. There was one CNC which was almost never used. It was not the right machine for their requirements.

In some instances there were constraints on interviewing conditions which were informative in themselves. Production managers or production engineering managers in Japan tended to hover close by during interviews of operators, and personnel managers were often present during interviews of union representatives. The latter were rather demure in the opinions they offered compared with their British counterparts. Vivid accounts of conflict, however, came from the personnel managers themselves, whose loyalty to the company was not in question, whereas some of the British personnel managers were concerned to present a harmonious picture of their factory. Although there was a greater reluctance to allow interviews of production workers during working hours in Japan – not so for managers –

in most cases the respondents were very generous with their time, and were open and helpful.

1 The Japanese factories

J1 is a small factory in Tokyo. The owner, his son and their families live above the factory and their wives are also on the payroll. J1 subcontracts for about 17 companies, but 25% of the work is done for its major contractor.[1] The family was under pressure to reduce costs by 30% in 1987, which they attempted to do by subcontracting out some finishing work and low value-added work, and cutting corners. The father runs the conventional machine shop, his son the CNC shop. An ex-employee works in a corner of the conventional machine shop on a lathe the father gave him. He turns parts for J1, more productively, it seems, than when he was an employee. With high land prices, the family may soon sell and move out of Tokyo.

J2 was established on the outer fringes of Tokyo when the owning family's Tokyo workshop was bombed during World War II. A family concern, it is run by a mother and three sons, and turns out high volume parts for video recorders, computers, theodolites, and so on, subcontracting for 10 companies in all. The main workshop whirrs with belt-driven machines (slotting, drilling, etc.) of pre-war vintage on one side, and automatic lathes on the other. Through into another room, and one is transported into the 1980s, with humming CNC machines, and hardly a worker to be seen. About half J2's employees are 'part timers' (which means that they do not work overtime on top of regular working hours, they are hourly paid, and their bonuses and 'retirement' money are considerably less than 'regular' workers. The part timers are all female except for one male retiree). Two youths work in the day time and go to high school in the evenings. With tightening margins for the low value-added work, controlling labour costs is a constant concern for the family.

J4 is also family-owned, but the present MD has eased other family members and relatives from active positions in the company. He is an ex-IBM employee, and took over the company at the death of his father, who was a widely respected, self-taught draughtsman and employed highly skilled craftsmen to produce J4's machines. The son attempted to restructure the flagging company by guaranteeing firm delivery dates and standardizing parts, and by bringing in young workers to rejuvenate the company. Most of the young workers are from a northern prefecture, and are being trained for the opening of a new branch factory there. The family owns two-thirds of the shares, and the other third is owned by other directors, of whom there are six, three who have been with the company for 30 years, and three who are recent arrivals. Four have a technical background. There is one part-time cleaning lady.

J9 had recently moved into its extremely spacious and modern premises, built at a cost of ¥2,000m. The president is an ex-employee of Nissan, which he left in 1960. He bought a scooter with his 'retirement money', his younger brother left Toshiba and bought a second-hand lathe with his, and together they formed the company. Other family members make up the rest of the directors but only these two have a technical background. The MD does sense grumblings sometimes that it's a 'family-run' company, but he says he is trying his best to educate his workers so that he can delegate more responsibility. They have always been involved with die (blanking die, special die, etc.) and always in hard materials like tungsten carbide, which their expertise gives them a competitive edge in. They are continually seeking to move towards higher value-added products. There are 8 non-regular (part-time or temporary) employees.

J45 is located near Tokyo, but its headquarters are in Tokyo itself. It manufactures a certain industrial machine, of which the government used to be a major purchaser, but it has been forced to seek new markets in recent years. The appreciation of the yen dealt a heavy blow to J45's plans to market its machines in the US, although it does export to SE Asia. The president is the largest shareholder (5.4%), followed by a trust bank (5%) and several other financial institutions. The employee shareholding association has just broken into the top ten shareholders.[2] There are 12 directors, four originally from the 'outside' (two from the principal bank, one ex-government official, one from a trading company). Three directors have technical backgrounds. Fifty-five of the workers are non-regular employees.

J50 is located south-west of Tokyo, but its headquarters are also in Tokyo. It makes industrial machines (cf. J66 and 4 of the British factories), which have a high engineering content, and batch sizes are very small. The industry has been depressed, which stimulated exploration of avenues for possible diversification. Only 5% of production is exported. The major shareholder is a major electrical and engineering group (13.2%), followed by a bank (4.3%), two insurance companies, another engineering company, and, sixth, the employees' shareholding association. There are 11 directors, mostly with long careers in the company, four with technical backgrounds and one with an accounting background. Sixty-four workers are non-regular employees. Most of these, as in the other factories, do office work rather than factory work.

J66 makes the same type of machine as J50, but is not a direct competitor. It exports 16–17% of total production. The recently deceased president was the major shareholder (12.4%) and a memorial fund he established the second largest (6.4%). The bank that installed him to rescue the company almost 40 years ago is third (4.9%); the employees' shareholding association ranks ninth. The former president's son is now the president. All six directors have been at J66 for at least 20 years: three have a technical background, one an accounting background. There are 69 part timers, plus some 20 loan (*shukko*) workers, on loan from a major customer in the shipbuilding industry hit by recession. The factory is carrying out a major capital investment programme, and had recently introduced TQC (total quality control) activities in preparation for its seventieth birthday in 1989.

J140 and **J180** belong to the same company, a company which has expanded extremely rapidly along with the company for which it makes parts (which is also the major shareholder at 22%). All its Japanese factories are within commuting proximity (it also has two foreign factories), and personnel and other policies are highly centralized. The second largest shareholder is another company from the same group (7.0%), a foreign company third, followed by three banks. The employees' shareholding association does not yet figure in the top 10. Of the 31 directors, one holds a concurrent outside post, the rest are evenly divided between those with engineering backgrounds and those with office-type backgrounds (four from accounting). Amongst other things, the company is famous for its vocational training programmes. Some 250 job descriptions form the basis of a kind of MBO programme for all employees. There are about 1,000 (mainly) temporary workers and part timers, as well as 600 'loan' workers who are the 'shock absorbers' to use the expression of one personnel manager.

2 The British factories

B4 once made a similar type of machine to J4, but the MD decided that future prospects were limited, and now the company sub-contracts anything from transfer tables to die; small-batch, high precision work, with the odd bit of designing. As in J4, the present MD is the son-in-law of the founder, who was also a gifted mechanical engineer. Again, the mother is the principal shareholder, and shares are held by four children besides the MD. Unlike J4, however, the number of employees has decreased over recent years, and there is a problem of young people leaving, particularly CNC operators. The 1980s have been a difficult decade, with rising costs (utilities, etc.) and decreasing profit margins.

B8 is one of four British factories producing the same type of industrial machine as J50 and J66. It was started by the director of an electronics company, but later sold to a foreign multinational when it was decided that it didn't fit the company's portfolio. A considerable amount of capital investment has since gone into the factory, which is now a specialist maker in the overall group. Low order books, however, coupled with technical difficulties with a new product resulted in a tenth of the workforce being made redundant in 1987. There are five directors; two based in Holland, the MD, the finance director, and an outside director who is a barrister. The multinational owner is seen as a generous parent giving considerable autonomy. There are three part timers (secretaries).

B11 makes the same kind of machine as B8, only smaller in size with a range of standard products, which are at the quality end of the market. B11 is reportedly still getting orders for parts for machines sold 40 years ago. The company is in its third generation of family ownership. The family has broader engineering interests, but B11 is the main concern. Two members of the family (non-technical backgrounds) are joint MDs, and there are now three other directors (two with technical backgrounds), two of whom were brought in around 1980, architects of the single-status initiatives which did a lot to reduce 'Us–Them' friction. There are no part timers.

B12 was started by a racing car driver to service his cars, but subcontracting work was taken on to fill in slack time. The same person still owns the company, and there are two directors with a smaller number of shares, one responsible for compressor assembling and the other for subcontracting – B12 – in factories adjacent to each other. The subcontracting director is very energetic. Since he came in 1981, he has sought to build stability and predictability in B12's operations, which has involved a strategy of turning out large batches for steady customers. These are done on a large number of CNC machines. He has convinced some of their contractors that reliability can be improved if they machine upstream operations as well. They have almost reached the design stage in some cases.

B39 was started by its US owner to service its machines (similar to those of J45) in Europe. Two years after start-up, there was much work in the US, and machine production was launched. The factory has been expanded several times, and some of the design work is done on site now, limited, though, because the US parent has its own R&D facility. Two of the three founders have gone back to the US, but their legacy, in personnel policies for example, remains. The remaining American is now the director of British operations. Seventy to eighty percent of production is exported. There are two part timers.

B71 had a long history as a family-owned engineering firm, traceable to the eighteenth century, but was bought by a US company in 1968 and is now 100% owned by its third set of US owners, a multinational with diverse interests. It is now officially a division of the group's UK company. B71 has become the standard products maker in the group of their type of machine, with limited scope for diversification. The MD is a member of the main board. There are seven local directors, with an average of 15 years at the factory, four with technical backgrounds, two with accounting backgrounds. The factory is a major employer in its rural location, and many managers and workers have been there for most if not all of their working lives. There are no part timers at the factory, but two at the London sales office.

B145 is the major factory of a well-known family concern which is now publicly listed. The group, which includes other engineering factories, has recently been expanding abroad. Although there has been a recovery in 1987–88, sales of B145 products have been hit hard by declining traditional markets and even newer markets, which is reflected in employment figures in table 1.2. There are nine directors, the top two of whom are also the top two of the group and five of whom have a technical background. About 50% of the machines are standard products, the rest engineered to varying degrees. Like B71, many families have been working at B145 for generations, but unlike B71, the factory is located in a traditionally strong trade union area. There are five part timers.

B80 and **B120** are also part of a publicly listed company, and have also suffered heavy retrenchments after losing ground to major international competitors (including J140/180). The various factories have considerable autonomy due to their geographical dispersal and to the fact that acquisition played a large part in past growth as against the almost total reliance on internal growth of J140/180. Major restructuring is being undertaken right across the group, however, with the introduction of modular production, QC-type activities, and so on. Half the directors of B80 have a technical background, with an average of 20 years in the company, and more than half of the directors of B120 have a technical background, with an average of 15 years. The company has also been noted for its training, of apprentices for example, and was the third company in the world to establish an off-job-training centre. Now retraining of workers has come to the fore with the large-scale restructuring effort. There are three part timers at both factories.

Notes

1 The new technology debate

1 See appendix 1 for technical terms and chapter 5 for the development of computerized machine tools.
2 Littler is probably unaware of the culturalist associations of the word *shudanshugi*; most of its proponents espouse a position he strongly denounces.
3 The word 'skill' is, of course, a vague term, as are 'deskilling' and 'reskilling'. It may refer to jobs, or workers, or both, or to capabilities which jobs are *supposed* to require or workers *supposed* to have. These are discussed later in the chapter.
4 See, for example, the ACARD Report (1983). Buchanan and Boddy (1984, 240) lament:

> The management aspirations that we appear to be looking at, and those we seem to be advocating are:

Actual	*Desired*
short term	long term
local, departmental	organizational
operating and control-oriented	strategy-oriented
technical	socio-technical
low risk	high performance

5 In the flexible firm jargon, Japanese firms are associated not only with functional flexibility, but also with numerical flexibility; the differentiation between a core of workers and various secondary groups which help insulate the core from disturbances caused by business fluctuations. The subtleties of these types of flexibility in Japan are often misunderstood, however.
6 See, for example, the papers presented to the Japanization of British Industry Conference, UWIST, 1987.
7 Keizai kikakucho (1986). Where the original source is English, the English name is used (e.g. Economic Planning Agency); where the source is Japanese, the Japanese name is used (e.g. Keizai kikakucho). Long vowel sounds in Japanese have not been distinguished from short vowel sounds. The authors of the OECD report *The Development of Industrial Relations Systems: Some Implications of the Japanese Experience* (1977) included a fourth pillar – social norms.

8 The masculine form is used throughout the book, perhaps excusably in this case because almost all the workers encountered were male. The few female shop-floor workers were usually doing lower-ranked jobs, which is worthy of a study in itself.

9 Shorn of their evolutionary connotations, Parson's pattern variables such as diffuseness versus specificity are also relevant, but OER does not represent a 'traditional' form of employment relations which are bound to 'modernize'.

10 See also Knights and Collinson (1985) for an example of low-trust initiatives being met with low-trust responses by workers in the context of job redesign.

11 See Becker (1964, ch.2). Human capital theory is often alluded to in Japan to explain differences with western countries, but the concepts are sometimes conflated; all skills formed in internal careers are deemed firm specific, hence general skills must be formed outside the company (see, for example, Koike, 1977, 5, 226) at the expense of the individual. It is debatable whether or not the *machining* skills of a lathe operator formed in a large Japanese factory – which are often reported to have multiskilling policy – are of a qualitatively different type from those of a British lathe operator who has changed company three times (and not all have), and whether those skills are not transferable. Significantly, Becker uses the example of machinists in his discussion of general training and skills. Most skills have a general and a firm-specific component. Often more than purely technical skills, firm specificity is related to the network of relationships formed inside and outside the company in the course of one's job, knowledge of physical layout and where different things are kept, different procedures for doing things, and so on (Aoki, 1984), but such firm-specific skills may also be sought after in the external labour market, as when someone is hired for their contacts.

12 A 'frontier of control' over the work process is meaningful where that control is contested, and it is less likely to be contested when it is not directly connected with the effort/reward equation – when, in other words, it is not the centre of the employment relationship, but one aspect of it (OER).

13 An example in the US is given by Thomas (1988, 19):

> The union, which had not been informed of the impending arrival of the CNC equipment by either management or the designated machine operators, filed an enquiry nearly as soon as the machines were bolted down. An extensive discussion with several shop stewards revealed two underlying concerns. The first concern was traditional: the new equipment altered the way operators worked (e.g., requiring them to manually input data) and the additional tasks seemed to warrant an increase in pay and, perhaps, a change in job classification. The existing classification – NC machine operator – was, they felt, inappropriate for the job since other workers with similar classifications were not trained to run the machines.

14 The survey covered 9,465 establishments of 50 employees or over in seven engineering industries (but the response rate was only 24.3%).

15 Social construction (of skill labels) theorists (e.g. Turner, 1962, Penn, 1982)

argue that there may be a large gap between both substantive skills and skill labels, and skill requirements and skill labels, as was shown when dilutees were brought into factories during the war time to carry out work that only 'skilled' men previously did. Restricting the numbers of people entering the trade is necessary to preserve skill labels hence pay differentials, they argue, and this is done through apprenticeships. Skill labels are obviously important when payment is based on the job done, or the amount of skill that a job is *supposed* to entail. It is in workers' interests in MER to keep skill labels as high as possible, and a large discrepancy between skill labels and skill requirements (in workers' favour) would indicate relatively strong worker power. Critics of the social constructionists, however, (e.g. More, 1982, Lee, 1982) argue that the defence of skill labels would hardly have been possible without employers actually needing the types of skills indicated by the labels, or as Dunlop (1964, 289) argues; 'technology and strategic position lead to apprenticed trades and not the reverse'. We would expect there to be little difference between skill labels and skill levels in OER, unless skill labels are inflated by managers for motivational reasons, since pay is not pegged directly to skills or the job done.

16 Kumazawa's (1970) work on deskilling precedes that of Braverman, and is based in part on a 1963 Nihon jinbungakkai study of technological innovation and its social consequences. He begins his article with the following quote from the *Nikkei shinbun*;

> The pressure to reduce costs in the bearing industry is severe, and prices in the past five years have more than halved. In order to achieve this, labour costs must be reduced and productivity increased. For this reason, the company adopted the fundamental policy in 1960 not to increase the number of employees beyond 200, and implemented a severe rationalization strategy. As a result the number of employees dropped in five years from 250 to 197, and monthly sales rose from ¥22–23m to ¥70m, achieving a 55% reduction in the cost of bearings. . . The company put women in shop-floor jobs in order to keep absolute labour costs down and decrease the proportion of labour in total costs. Female workers are limited to working 3–4 years, so the seniority-based wages can be kept in check. . . With female workers and standardization, simplification and specialization – especially important being automation and mechanization – a simple-work workshop was made. (6 Nov. 1965)

17 One large factory and one small one in each country making large batches (several hundred pieces on average) were selected to control for batch size. Five of the factories were 'large' (in principle more than *c.* 500 employees, with one exception in either country of similarly sized factories making similar products) and four 'small' (less than *c.* 100–120 employees). The fifth belonged to the same company as the large-batch, large factory in each country, and was intended to isolate further the influence of batch size, although the British factories had more independence from headquarters in terms of policy making and operations than

their Japanese counterparts. It is also useful to look at the factories as a spectrum through which the size effect can be traced.

Two large and two small factories in each country were to be making the same product. This was possible in Britain, but there were difficulties in interviewing operators in one smaller Japanese factory, and it only had two CNC machines. 'Anyway', said the production engineering manager, 'our CNC machines are largely for public relations – we don't use them very often.' Matchings by product-type are given in table 1.2.

18 Their employment relations are also reputedly different. These differences did not appear as great in Britain.

2 The wider context

1 The figures were obtained from the DE data base. The fact that few length of employment figures are published is of interest in itself, contrasting as it does with the amount of information on labour turnover.

2 According to the BWIR survey, the establishment is twice as likely to be the locus for decisions on wage increases at the company/divisional level (Millward and Stevens, 1986, 22). Brown (1981, 13) has noted that the choice is 'strongly affected by the heterogeneity of a company's products and by the history of its evolution by merger, takeover or internal growth', although he does seem to imply that the process is evolutionary towards the company level (1981, 11).

3 The policies of Zenkokukinzoku (National Federation of Metal and Engineering Workers) and Zenkindomei (Japanese Metal Industrial Workers Union), active in the industries looked at in this book, reflected the policies of their parent confederations Sohyo and the former Domei respectively.

4 Urabe (1984) criticizes Iwata's *shudanshugi* and other *nihonjinron* models for their assumptions of continuity. His account of the development of employment relations is useful, but he also develops his own *nihonjinron*, which contrasts cooperative Japanese employment practices with alleged rampant individualism in the west. Workers and managers view labour as a commodity in the latter, workers will change jobs at the whiff of better pay, and managers will lay workers off without batting an eyelid. General skills learned outside the enterprise predominate, individualism prevents QC circles from taking root, and so on.

5 The account given here is very short. Good English sources are Gordon (1985) for heavy industry, and Dore (1973).

6 According to Shirai (1986, 33–36), 7.6% of workers were organized in the period of 'Taisho democracy' in the late 1920s. The union movement was heavily suppressed following the Manchuria incident in 1931, and was further debilitated by internal ideological schisms, which were carried over into the post-war period.

7 Small and medium-sized enterprises in manufacturing are defined as those having less than 300 employees and capitalized at less than ¥100m. There is a Small and Medium Enterprise Agency which carries out research, publishes a

White Paper every year, and administers government policy directed at smaller companies.

8 Subcontracting in Japan developed soon after changes in labour markets and recession in the 1920s alluded to above produced wage discrepancies great enough to make outsourcing attractive, according to Nishiguchi (1989). Workers excluded from internal labour markets attempted to start their own businesses and by the 1930's, many had equipped themselves with power sources and primitive lathes and drilling machines.

Much subcontracting, however, was on a floating or spot deal basis (Watanabe, 1983a). The poor quality of subcontracted work was a problem, especially as imports dried up with the war. Companies started to move towards gradational – from trusted firms, often spinoffs – to those which could be used for 'spot' work – subcontracting. *Keiretsu*, or hierarchical groupings, were encouraged by the military government in its effort to rationalize resource allocation for weapons production.

After the initial post-war confusion, and with the return of demand in the 1950s, companies attempted to reestablish subcontracting relations. Dependency was created by the monopoly of imported technology by the larger companies (Watanabe, 1983), and subcontractors were hard pushed to keep up with technological change and orders from the 'parent' companies as mass production boomed. A number of developments, however, enabled some contractors to gain relative independence, including new financing options and increased technological expertise, which larger companies attempted to foster in striving to maintain or extend market share with product diversification (Nishiguchi, 1989, *passim*).

9 Contrasts, again, can be found in such industries as coal mining and construction (see De Vos and Mizushima, 1967).

10 The EPA (Keizai kikakucho, 1986, 134) gives the following figures for non-regular employees as percentages of the total workforce in 1985: part timers 6.4%; temporary, daily part timers 4.6%; temporary, daily general 5.4%.

11 The main sources for this account of Japanese corporate finance unless otherwise indicated are Kurosawa (1981, 84 and unpublished paper) and Okumura (1984, 1988 and unpublished paper).

12 Japanese companies also rely on external borrowing to a much greater degree than their British counterparts (Nihon Ginko, 1988, 73).

Again, ownership patterns have evolved from an early period of free markets, shaped by specific historical developments. Following the stock market crash of 1929 there was much discussion about reforming capital markets and war brought government intervention to ensure that financial resources were channelled into designated sectors. Capital markets were all but abolished in the late 1930s and banks became the main intermediaries for corporate financing. Although SCAP did attempt financial reforms such as reconstructing the bonds market after World War II, these attempts were cut short by changing priorities and strong capital market regulation was carried over. The following table is indicative of the changes.

Changes in external fund raising

	1931–36	37–45	46–55	56–65	66–75	76–78
Stocks, Ind. Bonds	96%	28%	18%	19%	10%	14%
Bank Loans	4	72	82	81	90	86

Source: Kurosawa, 1981 (originally in Nagatomi and Bank of Japan)

13 High debt/equity ratios result in demands for high risk compensation demanded by equity holders according to conventional 'market' wisdom. This does not seem to hold in Japan, however. In fact, the debt/equity distinction is somewhat blurred, as banks which loan money behave in some ways like equity holders, for example by rescheduling debts and involvement in corporate planning (see Corbett, 1987), and equity holders receive a very predictable dividend. Reasons behind the high price/earnings ratio are the subject of considerable debate. Okumura argues that mutual shareholding carried out for the reasons given above has raised the price of shares independent of dividends paid, and demand on the stock markets is more a function of capital gain prospects rather than dividends. He notes (1984, 61) that companies spend more on entertaining guests (*kosaihi*) than on total dividends to shareholders.

14 A resolution selected by delegates of the 1986 CBI annual conference but not discussed because of a shortage of time, for example, called on the conference to 'deplore the greed of the City and the lack of understanding displayed by those involved in providing capital to industry' (*Japan Times*, 13 November 1986).

15 Seven thousand managers of Matsushita Electric, for example, accepted 10% of their year-end bonuses in kind in 1986 as one of the company's measures to cope with the slump following the rise of the yen. Regular workers, presumably, received their normal bonuses (*Asahi Shinbun*, 8 November 1986). It is worth noting that the union presented the founder with a bronze statue worth ¥25m for his 92nd birthday in appreciation for his efforts at promoting cooperative industrial relations in the same month.

16 One must also recognize the existence of peer pressure as a mechanism of control.

17 Later in the study we shall see that in fact the amount of manual worker rotation (and promotion to non-manual jobs) was not so very different between many of the Japanese and British companies, but Japanese managers declared that they rotated workers for worker development and interest, and only after considerable probing did it emerge that 'production requirements' had a serious influence on the 'norm', whereas as one British manager remarked: 'A spade is a spade. We're moving them because of production requirements and tell them so.' He denied that they carried out rotation, despite the amount of movement between sections and departments.

18 The rapidity of ageing can be seen in the following. While it took 45 years (1930–1975) for the proportion of those aged 65 or above to move from 7% to 14% in

Britain, this will happen in 26 years in Japan (1970–1996; Rodosho, 1986, 10). Between 1973 and 1983 the average age of employees in large Japanese companies rose from 31.2 to 36.2 (Fujita, 1987, 57–8). Fujita estimates that an increase of five years can add 25% to wage bills in present terms.

19 For an account of Matsushita's wage system, see Matsushita denki sangyo rodo kumiai ed. (1986). According to a government survey 32.4% of those interviewed felt that *nenko joretsu* was a good system for workers, 28.7% felt it was good for both workers and companies, 5.6% felt it was good for companies, and 16.3% felt it was not a good system (Asahi Shinbun, 23 Nov., 1987).

20 For example, a conference was held at UWIST in 1987 under the title 'The Japanization of British Industry?'. See also Wood (1987), 'The Japanization of the Auto Industry?'

21 Another development in the factories in this study was the removal of leading hand and chargehand positions with the enlargement of operator responsibilities. This represented a trend *away* from Japanese practices, where there were several grades on the shop floor (see chapter 6) which were used to some degree for authority purposes, but more for training and promotion.

Japanese observers of British industrial relations are skeptical that recent trends represent fundamental changes. Ishida (1987), for example, argues that pay systems cannot motivate manual workers, who therefore work only to earn their daily bread. This in turn convinces those who design pay systems that that is the nature of the workers, hence the process is circular (c.f. Fox; see also White, 1981, 45). Unless this circularity in payment systems is tackled, he believes, there will be no fundamental change in British industrial relations.

The Industrial Participation Association also argues regarding share participation schemes that 'there is no way of providing satisfaction through a supplementary financial participation scheme if the wage structure itself is a source of discontent' (quoted in Grayson, 1984, 22).

22 Tom Mann Lecture quoted in *Personnel Management*, April 1986.

3 Employment relations 1

1 Recruitment is affected by a number of factors, not least amongst them, the economic 'climate'. Several of the British factories carried out large-scale redundancies in 1985 or 1986, and recruitment was severely curtailed. Thus the figures in table 3.1 are not a simple reflection of personnel policies, although they do give some indication of them.

2 The teacher in the northern provincial city sometimes arranged a package deal; 'These are the two you wanted, but please take this student as well.' He visited the factory from time to time to see what progress his ex-students were making. The recruits at J2 were *teijisei*, attending high school in the evenings. Some stayed after graduating, some were enticed off with the prospect of higher wages in other jobs (including driving rubbish trucks).

3 The manager added, 'Even if they don't stay with the company all their lives thirty years down the road they might discover the person on the other end of the

phone is a brother or sister apprentice.' B145 and B71 once had apprentice workshops, but had closed them down, and the first year of training (as well as administration of the aptitude tests) was done at the local polytechnic. Apprentice training is discussed in more detail in chapter 6.

4 Graduate engineers were one other group of British workers selected with extreme care and given a two- or three-day orientation course and extensive training. All of the factories which recruited graduate engineers, however, had trouble retaining them. 'If we do the milk rounds,' said the industrial relations manager at B71, 'We only keep them for two or three years.' His solution was to recruit apprentices in three streams, with all of them doing a common first year. Then the elites ('with 6 to 8 O levels') were sent to university, with the first year they had done counting as their EPI. Not only did they stay longer, he reasoned, but they received better training more suited for the company. This training was done in accordance with EITB regulations, as was the training of apprentices. As the figures indicate, however, retention rates were still low.

5 Although figures are incomplete, staff turnover seemed to be higher than that of manual workers – 4.4% as opposed to 3.3% at B145, for example.

6 The figures in table 3.2 do not indicate the distribution of ages and lengths of employment. The MD of J4 was trying to rejuvenate his workforce. There were the older workers from his father's time, and the young workers he had recruited. Without youth, he ventured, the company would not be able to adapt to technological innovation quickly enough. The age profile at B4 was similar, not because the MD was consciously trying to rejuvenate his workforce, but because he had found it very difficult to recruit workers for a number of years because of a nearby furniture store which paid much higher wages. It had closed down, and with high local unemployment, he had been able to recruit again.

7 This did not mean that they had spent their whole lives working within the firm, it must be added. Several had job-hopped on their way up. Some had come back to their old company, some had been recruited from outside.

8 The craft/technician distinction at B71 was relatively recent, which might inflate the figures, but the differences with the Japanese factories are clear.

9 Jones refers to the work of Koike (see, for example, Koike, 1988, 108, 123). Koike's main argument is that Japanese manual workers experience a greater range of jobs horizontally than do, for example, their US counterparts. He does note that the picture is different for machinists (1988, 142–3), but tends to generalize his steel and process industry findings.

10 The same manager also remarked, 'At the end of the year the organizational chart becomes a kind of matrix – people moving around everywhere, so they're used to doing different jobs. They don't think if they've been on lathes for 30 years that they shouldn't help out if there's a lot of work on milling. But people don't want too drastic changes, and I don't ask them to do things that are too different.' These remarks perhaps capture the subtlety of the mobility/stability issue in the factories.

11 Some of the Japanese factories hired agencies to check the backgrounds of their male university graduates, since some might become senior managers. The

practice of visiting homes of potential recruits was discontinued in some cases; the right to privacy was cited as the reason in J45. It was still done at J66 'to promote understanding both ways'.

12 Occasionally the factories were asked to take on trainee engineers and even apprentices of companies in difficulty (cf. *shukko* or 'loan' workers in Japan): 'We feel a kind of obligation, but once they finish they can stand on their own two feet' (B145).

13 Under the piecework system, workers apparently threw away defects, and even the blueprints with them.

14 Apprentices received the national minimum of £48.21 upon entry up to £100.49 for those 20 and over. Seventeen-year-old junior semi-skilled foundry workers received £56.68 while nineteen year olds received £85.56 – higher than the national minimums of £51.40 and £69.76 respectively.

15 In other words, production according to standard times divided by the actual time for that production, with different levels of bonus accruing to different percentages over 100%.

16 There were plans to have a 'single status canteen' when more money was available, but at the time of the interviews there were four sections.

17 The new works manager brought into B11 was the single status champion. Said the machine shop foreman, 'I used to have one big grievance – that the shop-floor were the dirt and the office were the cream. . . Things are a lot better now. It was the best thing they ever did in this firm.'

18 It was forced to give the differential after being taken to ACAS over war-time wage legislation (stating that the same rates had to be paid to workers doing the same job in factories involved in defence contracting) which was soon after repealed. The higher rate threatened the new wage system the company was about to introduce, but the problem was resolved when the company paid a lump sum and bought out the higher rate.

4 Employment relations 2

1 Part timers in the Japanese factories were not organized. Union executives had talked about it, but had made no serious moves in that direction, although the union at J140/180, for example, said that they did 'listen to their complaints'. The organization of lower managers introduces potential for role conflict. In one factory (not in this study) a department manager recalled how as a section chief (*kakaricho*) responsible to his section manager but also a union executive, he had taken his workers out on strike in the day time and smuggled himself and some workers back in at night so that the work would be done. There were other instances of *kakaricho*, who control the everyday work process to a large degree, ensuring that work could go on before taking their members out on strike. Did the workers resent this? 'No,' said the department manager, 'A lot of them are going to be *kakaricho* one day.'

2 The three levels of consultation were the *roshi kyogikai* (joint consultation meeting) held once a month to discuss business conditions and plans, attended

by company directors and senior union representatives; the *shokuba roshi kondankai* (workplace joint discussion meeting) also held once a month between chief stewards and factory and departmental managers to discuss company policy, wage and bonus prospects (the company side mainly reported in these meetings) and problems too big to be solved at the third and lowest level, the *shokuba kondankai* (workplace discussion meeting) also held once a month between union reps and *kakaricho* (section chiefs), in which some of the above topics were discussed, but more time was spent on issues relevant to the particular workplace, safety, and so on.

3 There were three workers on the works committee (council) at B8, all machinists. The chairman had recently gone, and no-one was keen to replace him, indicating apathy, the opposite of the involvement the MD was looking for. Two of the workers at B4 were elected in 1987 to go and talk to the MD: 'Things were getting slack, and people wanted to know what the situation was.' They promised the MD that they would not 'go running in with little problems', and that if there were grievances, they would go 'through the proper channels'. One of the reps was about to leave at the time of the interviews, and it was not sure that he would be replaced.

4 Income from union fees at J45 in 1986 was ¥15.4m, which would seem to place the figure at less than 1.6% of members' basic wages. The largest items of expenditure at the factory were 'activities' (*katsudo hi*, ¥4.6m), personnel (*jinkenhi*, ¥2.7m) and printing costs (¥1.3m).

5 The materials B120 produced were visually appealing, but the contents were simple, with no performance figures or analysis.

6 When communist party members gained control of the union, the company held a workshop (*gasshuku*) for managers and sent core workers on courses at the Japan Productivity Centre, but did not intervene directly, according to the personnel manager (he was also a former chairman of the union).

7 In 1950 there were major layoffs in many manufacturing industries, which generated intense industrial strife. While this marked the 'rollback' of union power, managers were anxious to avoid disputes arising from redundancies thereafter as far as possible.

8 A personnel manager at J140/180, where the term 'collective bargaining' was studiously avoided, had the following comments in this regard. First, even though there might only be a small margin in bargaining – which results not only from the company side but also from the influence of the national centres during the annual wage round – the union does push for the workers within that margin, and the company might give ground on other areas, or might give a hidden pay increase so as not to incur the wrath of other employers. The joint consultation meetings are sometimes very intense and, especially during worsening conditions, the union will fight hard for its members. With decreased room for manoeuvre in bargaining, other aspects of union activity have become increasingly important. Links with other unions in the industry give them solidarity which employers lack, so, for example, hours of work will be shortened through union efforts more than anything. They can also provide services for members

and their families, such as cheap loans, price reductions, lawyer services, and help in retirement planning. He concluded that if the union at J140/180 were strongly opposed to something, it would be very difficult for the company to act.

9 Although the unions insisted that all redundancies be voluntary, what actually happened was that letters were sent to 600 people the managers thought they would be prepared to lose stating that the company would be prepared to accept their resignation. Neither managers nor workers, however, thought that the company had used the opportunity to weed out militant unionists, although one worker did note that three out of five stewards in his department went. Moreover, according to one craftsman: 'The company wanted to get rid of dead wood, but they lost some very good workers, because they could still find jobs outside then.' Most of the workers were just thankful it was not them.

10 It would not be fair to describe the 'mood' at B145 and some of the other large engineering factories as being traditionally purely confrontational. B145 was a 'home away from home' where families had been working for two or three generations. That 'old caring family atmosphere' as one worker described it had been fading since the early 1970s, however, with an increase in younger, professional managers, the use of time and motion studies and so on. The personnel department was also sometimes associated with the demise of the 'old caring family atmosphere' with its formalization of industrial relations and the introduction of procedures. Despite these developments, layoffs and pressures from competition, these sentiments had not disappeared completely.

11 While this was welcomed in principle by the personnel and training managers, they were also cautious. Providing training for workers without openings to use their skills might lead to frustration and discord.

12 One survey of companies of over 100 employees found that 80% (presumably personnel managers) actively endorsed unions and most of the rest were passive assenters (Rodosho, 1987, 77).

5 Innovation

1 See appendix 1.

2 Nihon kosaku kikai kogyo kai (1988). The survey covered sales by the Association's 111 member companies. It would seem that wide-spread use of leasing has aided this spread. According to Mori (1982, 110), smaller firms overtook large firms in the purchase of NC/CNC in around 1972.

3 Even at J4, where the MD did consult the foreman briefly, he considered decisions regarding capital investment to be his sole responsibility.

4 A similar case of a small factory in Britain has subsequently come to my notice; the practice is not by any means unheard of in Britain.

5 In fact 'soft' is used to refer to even more diverse concepts. In an MOF series on 'Softnomics,' for instance, 'soft' is used to refer to anything from quality (as opposed to quantity) to invisible (as opposed to visible), information oriented, human oriented, and so on (Ishii *et al.*, 1985).

6 The slogan KKD to NTT is a play on the letters of the telecom companies KDD and NTT.

6 Training

1 There is not enough space here to go into detail on school education or public vocational training institutions. For these, see Prais, 1987, Dore and Sako, 1989, and Ishikawa, 1987.

2 In a factory studied by Batstone *et al.* (1987) the importance of craft was also reaffirmed with the introduction of CNC. Their account of the introduction of CNC parallels that of chapter 5 in a number of respects.

3 All the operators at B145 were operator/setters, but there was a division between grade 1 operating and grade 2 operating, the former referring to turning, universal grinding, universal milling and jig boring and the latter to other milling and drilling work. Two operators had been upgraded from grade 2 to grade 1, even though they were not time-served.

4 The company paper in B71 carried an article in late 1985:

> Many congratulations to X and Y on completing 50 years active service with the Company. It is a remarkable achievement for any standard to work for one employer for 50 years. It is even more remarkable to do the whole of that time in the same department as they both have.

The machine shop superintendent commented that people would no longer do the same job for 50 years. Young operators wanted to move about and experience different jobs, and also to progress upwards. Two of the older CNC operators – one the outgoing senior steward – said they could have taken an office job but declined. Nonetheless, they had done a wide range of jobs in the machine shop; apprenticeship, boring section (5–6 yrs), tool room (7 yrs) and 5 different CNC machines, including machining centres (10 yrs) for one; the steward said he had worked on most manual machines, and NC and CNC for the last 16 years.

Two long serving operators in J45 who went through the company school had had the following careers; deburring and other odd jobs at first, followed by 15 years on milling machines, thence on to a machining centre for one; and 14 years on lathes, 3 years on an NC lathe, thence over to milling (one year on a manual milling machine and 5 years on a CNC miller) and on to a machining centre for the other.

5 As the following discussion with Koseki Tomohiro (lathe operator and author of several books about small factories) suggests regarding the craft mentality in small factories:

> Ten years ago only I could use – and the MD knew – CNC. We had one CNC lathe. There were 20 skilled workers (craftsmen) there then. Within five years, machining centres came in. Now half of the skilled workers have left. Our age is past, they thought. One started plumbing, one went back to farming. They weren't even laid off, but in smaller factories they don't get other jobs, and they leave by themselves. Even if you gave them the same wages for doing a simple job they'd leave – their pride wouldn't let them do that like they might in big factories.

6 Regarding OJT, Ishikawa has the following observation:

> During the period of high economic growth, many production plants recruited youths who completed the compulsory education and provided intensive craft training off-the-job. This traineeship was followed by planned OJT [on-the-job training] and the traineeship lasted for three years. This was similar in outlook to the Western apprenticeship, but differed significantly in its functioning. The training did not normally lead to any recognized occupational qualification, nor social status, but formed a basis on which the trainees were expected to build additional skills within the firm. . . The duration of this initial off-JT has been reduced as the recruitment source has shifted to senior high school graduates with 12 years' schooling, and the learning of skills has come to be effected more on the job supplemented by occasional off-JT. Employers generally consider OJT as most important, *although planned OJT is not very well developed in practice* (Ishikawa, 1987, 35, emphasis added)

7 The principal objective of the government in establishing the tests, according to Ishikawa, was to raise the social status of blue collar workers by giving recognition to their skill levels, and to motivate workers towards skill acquisition (1987, 18).

8 The factory manager at J2 also said: 'They teach computing at most high schools these days, so it would only take a week or two to replace an operator who left.' The setting work at J2 was relatively simple, however (see chapter 7).

9 The difference in retirement ages must also be considered, but it was not the most important factor. Nor was the supply of young operators *per se*, for more young workers just finishing their apprenticeships or not having done an apprenticeship might have been used in the British factories.

10 The industrial relations manager wondered if in fact this was not too young, because operators in whom they had invested a lot of time and money were more likely to move if they were under 30.

11 At J50 and J66 there were about five in a group as well as two levels of foreman and a superintendent level. There was a group leader and at least one (technical) 'trouble shooter' for 8–10 workers at J140 and J180, and three to four groups under a foreman.

12 According to one source, there were 3,300 microelectronics-related mechanical courses offered at public vocational training institutions in 1986, 720 of which could be considered upgrading or retraining courses, but what proportion of the students were sent on the courses by employers, or were employed during that time, was not indicated (Rodosho shokugyo noryoku kaihatsu kyoku, 1987). Public training facilities have tended to carry the stigma of being places for retraining unemployed or people who haven't been able to get regular jobs rather than places where companies might send their employees for courses (Ishikawa, 1981, 32).

13 J50 also sends about 10 workers a year – usually group leaders or foremen – to the Central Skill Development Centre in Chiba to absorb new ideas, but none of the courses were related to CNC.

7 Division of labour

1 On conventional machines, it is difficult for the operator to operate in more than one dimension simultaneously, which results in many special fixtures and increases the difficulty of setting up. With CNC, which can machine in several dimensions at once, more standard fixtures can be used.

2 If programming is done in an office, it is fed into the machine controller by such means as a tape or portable hard disc, or even directly from machine to machine by a process called direct numerical control (DNC). As mentioned in chapter 5, with the development of CNC, alterations could be done directly at the machine controller, and with manual data input (MDI), all programming can be done through the keys on the control panel. This can be done while a part is being machined with background programming.

3 He had pushed for MDI before he became manager: 'I sat down with J (production-engineering manager) and Y (works manager) and we were asked what we would go for. There was disagreement, and finally we went for my suggestion with the understanding that it was on my head. It worked well. There were great reservations about MDI but within a month we were asking for another one.'

4 Walking around the shop-floor with the production engineering manager, I came across a programmer doing his own programming. Said the production engineering manager: 'Yes, he doesn't have a very professional attitude, but he can do it. He does one offs, and it takes a bit of a load off the programmers.'

5 This was tantamount to saying it was a low status, low skill job.

6 Factory size and batch size are multiplied, divided into two groups and correlation tested with task range.

7 Babbage considered Adam Smith's concept of the division of labour incomplete and, writing in 1835 (Babbage, 1971), added what is now called the Babbage Principle, which involves decomposing a skilled job into an essential skilled component (which is still done by a craftsman), and peripheral tasks which need not be done by a craftsman.

8 CNC use and skills

1 Experience being used in a negative sense here; relying on methods that have always been used without questioning the logical principles behind them, or whether they could be done better.

2 There is, however, no shortage of warnings about the shortsightedness of losing conventional machining skills, particularly from researchers. One survey by the Koyo shokugyo sogo kenkyujo showed that managers and workers do think intuition and experience are still important. In view of the findings of this study, these arguments, plus arguments that traditional skills are still necessary, are akin to normative statements about mobility and job rotation. Slogans emphasizing a shift in skills may coexist with exhortations to keep manual skills. Like rotation and internal mobility, such practices as multiskilling are interpreted (by the researchers, but the interpretation is fed back into industry) as being

something peculiarly Japanese (eg Shokugyo kunren kenkyu senta, 1983, 22–3).
3 Interview 12 February 1987.
4 Some managers, as noted, foresaw the day that skilled operators would ask to be taken off CNC, but they were not going to initiate that process. Significantly for the following discussion, there was no suggestion that they were to be used as vanguards for more advanced forms of automation. Presumably they would be transferred back to manual machines.
5 Interview 12 February 1987.
6 Totsuka *et al.* studied industrial relations in the British automobile and steel industries, mainly in the late 1970s, but they suggest that the degree of management involvement had not significantly changed by the early 1980s.
7 In his study of FMS systems in Japan, Jaikumar (1986) found small teams of engineers, a high proportion of which were graduates, given responsibility for day-to-day operations until uptime of 90–95% (unattended) had been achieved. In the factories in this study, too, production engineers were responsible for the newest technology until it was debugged and running routinely, hence 'technology transfer' associated with setting up was not as great as it might have been.
8 Strategic control of information is a widely recognized phenomenon not limited to production engineers, of course, but employment relations in Japanese factories can be said to reduce the incentives, both in terms of pecuniary gain and job security.

Appendix 1

1 See Leatham Jones (1986), Gibbs (1984), and Trent (1984).

Appendix 2

1 J1 had obtained a verbal promise from the contractor that it would not purchase the same kind of CNC machines before it invested in (actually leased) its own.
2 It was started in 1975 amid the general fear of companies falling into the hands of speculator (*seibi*) groups. The employees were supposed to be additional 'safe' shareholders. The scheme is optional; the manager referred to pays ¥5,000 per month into the scheme and ¥15,000 at bonus times. The buying of shares is done by a trust bank. Unlike some other employee shareholder schemes, the employees can only sell the shares when they leave the company. The scheme is not, apparently, considered a major motivational instrument.

Bibliography

Abegglen, J. (1985), *The Japanese Factory*, Glencoe IL: The Free Press.

Abegglen, J., and G. Stalk (1985), *Kaisha: The Japanese Corporation*, New York: Basic Books.

ACARD (1983), *New Opportunities in Manufacturing*, London: HMSO.

Aldridge, A. (1976), *Power, Authority and Restrictive Practices: A Sociological Essay on Industrial Relations*, Oxford: Basil Blackwell.

Aoki, M. (1984), *The Cooperative Game Theory of the Firm*, Oxford: Oxford University Press.

Arnold, E. (1983), 'Information Technology as a Technological Fix' in G. Winch, ed., *Information Technology in Manufacturing Processes*, London: Rossendale.

Arthurs, A. (1985), 'Towards Single Status?' *Journal of General Management*, Vol. 11, No. 1.

Babbage, C. (1971), *On the Economy of Machinery and Manufactures* (1835), Fairfield NJ: Kelley.

Bain, G., ed. (1983), *Industrial Relations in Britain*, Oxford: Basil Blackwell.

Ballon, R. (1986), *Labour–Management Relations in Japan*, Tokyo: Sophia University.

ed. (1969), *The Japanese Employee*, Tokyo: Sophia University.

Batstone, E. (1988), *The Reform of Workplace Industrial Relations*, Oxford: Clarendon Press.

Batstone, E., S. Gourlay, H. Levie and R. Moore (1987), *New Technology and the Process of Labour Regulation*, Oxford: Clarendon Press.

Becker, G. (1980), 'Altruism in the Family and Selfishness in the Market Place', Discussion Paper, London School of Economics.

(1964), *Human Capital*, New York: Columbia University.

Berger, S. and M. Piore (1980), *Dualism and Discontinuity in Industrial Societies*, Cambridge MA: Harvard University Press.

Bessant, J. (1983), 'Management and Manufacturing Innovation: Information Technology' in G. Winch ed. *Information Technology in Manufacturing Processes*, London: Rossendale.

Braverman, H. (1974), *Labor and Monopoly Capital*, New York: Monthly Review Press.

Brown, Wilfred (1962), *Piecework Abandoned*, London: Heinemann.

Brown, William (1986), 'The Changing Role of Trade Unions in the Management of Labour', *British Journal of Industrial Relations*, Vol. 24, No. 2, July.
(1981), *The Changing Contours of British Industrial Relations*, Oxford: Basil Blackwell.
Buchanan, D. (1983), 'Technological Imperatives and Strategic Choice' in G. Winch ed. *Information Technology in Manufacturing Processes*, London: Rossendale.
Buchanan, D., and D. Boddy eds. (1983), *Organisations in the Computer Age*, Aldershot: Gower.
Child, J. (1984), *Organisations: A Guide to Problems and Practice*, 2nd edition, London: Harper and Row.
Child, J., and M. Tayeb (1982–3), 'Theoretical Perspectives in Cross-National Organizational Research', *International Studies of Management and Organization*, Vol. 12, No. 4, Winter.
Chusho kigyo cho (1986), *Chusho kigyo hakusho* (White Paper on Small and Medium-Sized Enterprises), Tokyo: Okurasho.
(1985), *Haiteku kiki donyu handobukku: Chusho kigyo setsubi kindaika no tebiki* (Handbook for Introducing High-tech. Machines: Manual for Modernization of Equipment in Small and Medium-Sized Enterprises), Tokyo: Tsusho sangyo chosa kai.
Clark, R. (1979), *The Japanese Company*, New Haven: Yale University Press.
Clegg, H. (1979), *The Changing System of Industrial Relations in Britain*, Oxford: Basil Blackwell.
(1976), *Trade Unionism Under Collective Bargaining*, Oxford: Basil Blackwell.
Cole, R. (1979), *Work, Mobility and Participation: A Comparative Study of American and Japanese Industry*, Berkeley: University of California Press.
(1971), *Japanese Blue Collar: the Changing Tradition*, Berkeley: University of California Press.
Corbett, J. (1987), 'International Perspectives on Financing: Evidence from Japan', *Oxford Review of Economic Policy*, Vol. 3, No. 4.
Cross, M. (1985), *Towards the Flexible Craftsman*, London: Technical Change Centre.
Daiyamondo sha ed. (1980), *Keiei jitsumu daihyakka II* (Encyclopedia of Practical Business Management II), Tokyo.
Daniel, W. (1987), *Workplace Industrial Relations and Technical Change*, London: PSI and Frances Pinter.
(1978), 'The Effects of the Employment Protection Laws in Manufacturing Industry', *Employment Gazette*, Vol. 86, No. 6.
Department of Employment (1986), *New Earnings Survey*, London: HMSO.
De Vos, G., and K. Mizushima (1967), 'Organization and Social Function of Japanese Gangs: Historical Development and Modern Parallels', in R. Dore ed., *Aspects of Social Change in Modern Japan*, Princeton: Princeton University Press.
Dobson, C. (1980), *Masters and Journeymen: a Prehistory of Industrial Relations, 1717–1800*, London: Croom Helm.
Dodgson, M. (1985a), *Advanced Manufacturing Technology in the Small Firm*, London: Technical Change Centre.

(1985b), 'Work Organization and Skills in Small Engineering Firms Using CNC Machine Tools', Ph.D. dissertation, Imperial College, London.
Doeringer, P., and M. Piore (1971), *Internal Labour Markets and Manpower Analysis*, Lexington MA: D.C. Heath and Co.
Donovan, Royal Commission on Trade Unions and Employers' Associations (1986), *Report*, London: HMSO.
Dore, R. (1987), *Taking Japan Seriously*, London: Athlone Press.
(1985a), 'Financial Structures and the Long-Term View', *Policy Studies*, Vol. 6, Part 1, July.
(1985b), 'The Confucian Recipe for Industrial Success', *Government and Opposition*, Vol. 20, No. 2, Spring.
(1984), 'Niju kozo no saikento' (Reconsidering Dual Structure), *Nihon Rodo Kyokai Zasshi* (Journal of the JIL), April/May.
(1983), 'The Social Sources of the Will to Innovate', public lecture at Imperial College: Imperial College/SPRU/TCC.
(1979), 'Industrial Relations in Japan and Elsewhere' in A. Craig ed. *Japan – a Comparative View*, New Jersey: Princeton University Press.
(1973) *British Factory – Japanese Factory*, Berkeley: University of California Press.
Dore, R., and M. Sako (1987), 'Vocational Education and Training in Japan', report for Manpower Services Commission.
Dunlop, J. (1964), 'Review of Turner', *British Journal of Industrial Relations*, 14, 2.
Economic Planning Agency, see Keizai kikakucho.
Edwards, P. and H. Scullion (1982), *The Social Organization of Industrial Conflict: Control and Resistance in the Workplace*, Oxford: Basil Blackwell.
Edwards, R. (1979), *Contested Terrain: the Transformation of the Workplace in the Twentieth Century*, London: Heinemann.
Elger, T. (1982), 'Braverman, Capital Accumulation and Deskilling' in S. Wood ed. *The Degradation of Work*, London: Hutchinson.
Elliot, R., and J. Fallick (1981), *Pay in the Public Sector*, London: Macmillan.
Etzioni, A. (1961), *A Comparative Analysis of Complex Organizations*, Glencoe: Free Press.
Foucault, M. (1976), *The History of Sexuality*, Harmondsworth: Penguin, 1981.
Fox, A. (1985a), *History and Heritage*, London: Allen and Unwin.
(1985b), *Man Mismanagement*, 2nd edition, London: Hutchinson.
(1974), *Beyond Contract: Work, Power and Trust Relations*, London: Faber and Faber.
Francis, A. (1986), *New Technology at Work*, Oxford: Oxford University Press.
(1976), 'Families, Firms and Finance Capital', unpublished working paper, Nuffield College.
Francis, A., J. Turk and P. William (1983), *Power, Efficiency and Institutions*, London: Heinemann.
Friedman, A. (1977), *Industry and Labour: Class Struggle at Work and Monopoly Capital*, London: Macmillan Press.
Friedman, D. (1988), *The Misunderstood Miracle: Industrial Development and Political Change in Japan*, Ithaca: Cornell University Press.

Fujino, M. (1982), 'Shokuba no ME kakumei to rodo kumiai no han'o' (The Microelectronics Revolution in the Workplace and the Response of Labour Unions) in *Saisentan gijutsu to soshiki, ningen* (Advanced Technology and Organizations, People), Tokyo: Nihon Seisansei Honbu.

Fujita, Y. (1987), 'Koreika mondai' (The Aging Problem), in Nikkeiren ed., *10 nen go no jinji romu* (Personnel management in 10 years), Tokyo.

Gallie, D. (1978), *In Search of the New Working Class*, Cambridge: Cambridge University Press.

Gennard, J. (1976), 'Multinationals–Industrial Relations and Trade Union Response, Occasional Paper, Universities of Leeds and Nottingham.

Gibbs, D. (1984), *An Introduction to CNC Machining*, London: Cassell.

Gluck, C. (1985), *Japan's Modern Myths: Ideology in the Late Meiji Period*, Princeton NJ: Princeton University Press.

Gordon, A. (1985), *The Evolution of Labour Relations in Japan: Heavy Industry, 1853–1955*, Cambridge MA: Harvard University Press.

Gould, W. (1984), *Japan's Reshaping of American Labour Law*, Cambridge MA: MIT Press.

Grayson, D. (1984), 'Progressive Payment Systems', Occasional Paper 28, Work Research Unit, London: ACAS.

Halevi, G. (1980), *The Role of Computers in Manufacturing Processes*, New York: John Wiley and Sons.

Hanami, T. (1983), 'The Function of Law in Japanese Industrial Relations' in T. Shirai ed. *Contemporary Industrial Relations in Japan*, Madison WI: University of Wisconsin Press.

(1979), *Labour Relations in Japan Today*, Tokyo: Kodansha International.

Hattori, T. (1986), 'Technology Transfer and Management Systems', *The Developing Economies*, Vol. 24, No. 4, December.

Hiraishi, N. (1987), *Social Security*, Tokyo: Japan Institute of Labour (Japanese Industrial Relations Series).

Hobsbawm, E. (1984) *World of Labour*, London: Weidenfeld and Nicolson.

(1968), *Labouring Men*, 2nd edition, London: Weidenfeld and Nicolson.

Hyman, R. (1975) *Industrial Relations: A Marxist Introduction*, London: Macmillan.

Hyman, R., and Brough, I. (1983), 'Trade Unions: Structure, Policies and Politics' in G. Bain ed. *Industrial Relations in Britain*, Oxford: Basil Blackwell.

(1975), *Social Values and Industrial Relations: A Study of Fairness and Inequality*, Oxford: Basil Blackwell.

Inaba, S. ed. (1970), *Yasashii NC tokuhon* (Easy NC Reader), Tokyo: Nihon Seisansei Honbu.

Inagami, T. (1989), *Tenkanki no rodo sekai* (World of Labour in Transition), Tokyo: Yushindo.

(1983), 'Niyu tekunoroji to rodo kumiai' (New Technology and Labour Unions) in *Nihon rodo kyokai zasshi* (Journal of the JIL), Vol. 28, No. 9, October.

(1981), *Roshi kankei no shakaigaku* (The Sociology of Industrial Relations), Tokyo: Kenkyusha.

Inagami, T., and T. Kawakita, eds. (1988), *Yunion aidentetei* (Union Identity), Tokyo: Nihon Rodo Kyokai.
Ingham, G. (1974), *Strikes and Industrial Conflict*, London: Macmillan.
(1970), *Size of Industrial Organisation and Worker Behaviour*, Cambridge: Cambridge University Press.
Inohara, H. (1985), *The Japanese Personnel Department: Structure and Functions*, Tokyo: Sophia University.
Ishida, M. (1987), 'Igirisu chingin seido yori mita roshi kankei no genzai' (Present Industrial Relations as seen through British Wage Systems), in *Nihon Rodo Kyokai Zasshi* (Journal of the JIL), Feb./March.
Ishii, T., *et al.* (1985), *Sentan gijutsu no kakushin to sono eikyo* (Advanced Technological Innovation and Its Influence), Tokyo: Okurasho.
Ishikawa, T. (1987), *Vocational Training*, Tokyo: Japan Institute of Labour (Japanese Industrial Relations Series).
Itami, H. (1987), *Jinbonshugi kigyo* ('Peopleism' Enterprises), Tokyo: Chikuma shobo.
Itami, T., and Y. Matsunaga (1985), 'Chukan rodo shijo ron' (Theory of Intermediate Labour Markets) in *Nihon rodo kyokai zasshi* (Journal of the JIL), May.
Jaikumar, R. (1986), 'Post-industrial Manufacturing', *Harvard Business Review*, November–December.
Jamieson, I. (1980), *Capitalism and Culture: a Comparative Analysis of British and American Manufacturing Organizations*, Aldershot: Gower.
Japan External Trade Organization (JETRO) ed. (1985), *Japan's Postwar Industrial Policy*, Tokyo.
Japan Institute of Labour (JIL) ed. (1984), *Microelectronics and the Response of Labour Unions*, Tokyo.
Japan Labour Bulletin, various issues.
See also *Nihon rodo kyokai.*
Japan Machine Tool Builders' Association ed. (1985), *Machine Tool Industry Japan 1985*, Tokyo.
See also *Nihon kosaku kikai kogyokai.*
Japan Productivity Centre (1986), *Productivity Movement in Japan*, Tokyo.
Johnson, C. (1982), *MITI and the Japanese Economic Miracle: the Growth of Industrial Policy 1925–75*, California: Stanford University Press.
Jones, B. (1986), 'Cultures, Strategies and Technical Essentials: a Comparative View of Work and Flexible Production Technology', paper presented to the International Workshop on New Technology and New Forms of Work Organization, Berlin.
(1982), 'Destruction and Redistribution of Engineering Skills?' in S. Wood, ed. *The Degradation of Work?* London: Hutchinson.
Joyce, P. (1980), *Work, Society and Politics: the Culture of the Factory in Later Victorian England*, Brighton: Harvester Press.
Kahn-Freund, O. (1954), 'Intergroup Conflicts and their Settlement', *British Journal of Sociology*, Vol. 5, No. 3, September.
Kameyama, N. (1983), 'ME gijutsu kakushin to koyo mondai' (Microelectronics

Technological Innovation and Employment Problems) in T. Kuroda ed. *ME kakumei to rodo kumiai* (The Microelectronics Revolution and Labour Unions), Tokyo: Nihon Hyoronsha.

Kaplinsky, R. (1984), *Automation: the Technology and Society*, Harlow: Longman.

Keizai kikakucho ed. (1987), *Shokugyo kozo henkanki no jinzai kaihatsu* (Development of Human Resources in a Period of Changing Occupational Structures), Tokyo: Okurasho.

(1986), *Gijutsu kakushin to koyo* (Technological Innovation and Employment), Tokyo: Okurasho.

Kelley, M. (1988), 'Beyond the Deskilling Debate: A Statistical Model for Explaining the Design of Blue Collar Jobs Under Programmable Automation', Harvard University, Kennedy School of Government.

Kelley, M., and H. Brooks (1988), 'The State of Computerized Automation in U.S. Manufacturing', Harvard University, Kennedy School of Government.

Kigyo katsuryoku kenkyujo ed. (1986), *Shushin koyo seido no doko ni kansuru chosakenkyu* (Survey Research on Trends in the Lifetime Employment System), Tokyo.

Kilpatrick, A., and T. Lawson (1980), 'On the Nature of Industrial Decline in the UK', *Cambridge Journal of Economics*, IV.

Knights, D., H. Willmott and D. Collinson eds. (1985), *Job Redesign: Critical Perspectives on the Labour Process*, Aldershot: Gower.

Koike, K. (1988), *Understanding Industrial Relations in Modern Japan*, London: Macmillan.

(1983), 'Workers in Small Firms and Women in Industry' and 'Internal Labour Markets: Workers in Large Firms' in Shirai ed., *Contemporary Industrial Relations in Japan*, Madison WI: University of Wisconsin.

(1981a), *Chusho kigyo no jukuren* (Skills in Small and Medium-Sized Enterprises), Tokyo: Dobunkan.

(1981b), *Nihon no jukuren: sugureta jinzai keisei shisutemu* (Skills in Japan: System of Outstanding Human Resource Formation), Tokyo: Yuhikaku.

(1977), *Shokuba no rodo kumiai to sanka* (Workplace Labour Unions and Participation), Tokyo: Toyo keizai shinbun sha.

Koseki, T. (1986), *Machi koba no jikai* (The Magnetic World of the Street-Corner Workshop), Tokyo: Gendai shokan.

Koshiro, K. (1986a), 'Labour Market Flexibility in Japan – With Special Reference to Wage Flexibility', Discussion Paper, Yokohama National University, Faculty of Economics.

(1986b), *see* Ujihara

(1983), 'Labour Relations in Public Enterprises' in T. Shirai ed. *Contemporary Industrial Relations in Japan*, Madison WI: University of Wisconsin Press.

(1979), *Labour-Management Relations in Japanese Public Enterprises*, Berlin: Studienverlag Brockmeyer.

Koyo shokugyo sogo kenkyujo ed. (1985a), *ME gijutsu kakushin no genba rodosha ni oyobosu eikyo* (The Effects of Microelectronics Innovation on Shopfloor Workers), Tokyo.

(1985b), *Maikurerekutoronikusuka to seisan gijutsu, shokuba soshiki no henka ni*

kansuru kenkyu hokokusho (Research Report on Microelectronicization and Changes in Production Engineering and Workplace Organization), Tokyo.
(1983), *ME no koyo ni oyobosu shitsuteki eikyo ni kansuru kenkyu hokokusho* (Research Report on the Qualitative Effects of Microelectronics on Employment), Tokyo.
See also *National Institute of Employment and Vocational Research* (NIEVR)
Kumazawa, M. (1986) *Shokuba no shura o ikite* (Living the Workplace Carnage), Tokyo: Chikuma Shobo.
(1970), 'Rodo tanjunka no ronri to genjitsu' (Theory and Reality of the Simplification of Labour), *Nihon rodo kyokai zasshi* (Journal of the JIL), June.
Kuroda, T. ed. (1983), *ME kakumei to rodo kumiai* (The Microelectronics Revolution and Labour Unions), Tokyo: Nihon Hyoronsha.
Kurosawa, Y. (1984), 'Business Risk, Dividend Policy and Policy for Capital Structure: an Empirical Study of Japanese Enterprises', Japan Development Bank Staff Papers.
(1981), 'Corporate Financing and Capital Markets', Japan Development Bank Staff Papers.
Kuwahara, Y. (1987), 'Organization and Policy of Labour Unions in the Advanced Technetronics Age', paper presented to International Conference on Labour Problems in the Advanced Technetronics Age, Tokyo.
(1985), 'Labour and Management Views of and Their Responses to Microelectronics in Japan', paper presented to International Conference on Microelectronics and Labour, Tokyo.
(1983), 'Technical Change and Industrial Relations in Japan', *Bulletin of Comparative Labour Relations* Bulletin 12, Special Issue.
Leatham-Jones, B. (1986), *Introduction to Computer Numerical Control*, London: Pitman.
Lee, D. (1982), 'Beyond Deskilling, Craft and Class' in S. Wood ed. *The Degradation of Work?* London: Hutchinson.
(1981), 'Skill, Craft and Class: a Theoretical Critique and a Critical Case', *Sociology*, Vol. 15.
Leeson, R. (1979), *Travelling Brothers: the Six Centuries Road from Craft Fellowship to Trade Unionism*, London: Allen and Unwin.
Legge, K. (1978), *Power, Innovation and Problem Solving in Personnel Management*, London: McGraw-Hill.
Levine, S. (1984), 'Careers and Labour Mobility in Japanese Labour Markets' in D. Plath ed. *Work and Lifecourse in Japan*, Albany NY: SUNY.
Levine, S., and H. Kawada (1980) *Human Resources in Japanese Industrial Development*, Princeton NJ: Princeton University Press.
Littler, C. (1982), *The Development of the Labour Process in Capitalist Societies*, London: Heinemann.
Littler, C., and G. Salaman (1984), *Class at Work: the Design, Allocation and Control of Jobs*, London: Batsford.
Lloyd, P., and R. Blackwell (1983?), 'A Review of Substantive Issues in Civil Service Industrial Relations Since the Early 1970s', unpublished report.
Lukes, S. (1974), *Power: a Radical View*, London: Macmillan.

Macfarlane, A. (1978), *The Origins of English Individualism*, Oxford: Basil Blackwell.

McGregor, D. (1960), *The Human Side of Enterprise*, New York: McGraw-Hill.

Manwairing, T. (1981), 'The Trade Union Response to New Technology', *Industrial Relations Journal*, Vol. 12, No. 4.

Martin, R. (1981), *New Technology and Industrial Relations in Fleet Street*, Oxford: Oxford University Press.

(1977), *The Sociology of Power*, London: Routledge and Kegan Paul.

Maruo, N. (1986), 'Nihongata keiei to seika haibun' (Japanese-style Management and the Distribution of Profits), *Nihon Rodo kyokai zasshi* (Journal of the JIL), Vol. 28, No. 9, September.

Marx, K. (1976), *Capital, Vol. 1*, Harmondsworth: Penguin.

Matsushita denki sangyo rodo kumiai ed. (1986), *Matsushita denki no shin shigotobetsu chingin* (Matsushita Electric's New Job-based Wages), Tokyo: Sangyo rodo kenkyujo.

Maurice, M. (1985), 'Microelectronics and Changes in Job Content, Job Requirements and Sequence of Job Promotion or Workers' Careers', paper presented to International Conference on Microelectronics and Labour, Tokyo.

Maurice, M., A. Sorge and M. Warner (1980), 'Societal Differences in Manufacturing Organizations: a Comparison of France, West Germany and Great Britain', *Organization Studies*, Vol. 1, No. 1.

Mihara, Y. (1986), 'Kikai kako shokuba no gino keisei' (Skill Formation in Machine shops), in Koike, ed., *Gendai no jinzaikeisei* (Contemporary Human Resource Formation), Kyoto: Minerubua Shobo.

Millward, N., and M. Stevens (1986), *British Workplace Industrial Relations*, Aldershot: Gower.

Ministry of Labour (Japan), see *Rodosho*.

MITI, see *Tsusansho*.

Moore, J. (1983), *Japanese Workers and the Struggle for Power 1945–47*, Madison WI: University of Wisconsin Press.

More, C. (1982), 'Skill and the Survival of Apprenticeship', in Wood ed., *The Degradation of Work?*, London: Hutchinson.

Mori, K. (1982), *Machi koba no roboto kakumei* (The Robot Revolution in Street-Corner Workshops), Tokyo: Daiyamondosha.

Morris, T. (1986), *Innovations in Banking: Business Strategies and Employee Relations*, London: Croom Helm.

Mouer, R., and Y. Sugimoto (1986), *Images of Japanese Society*, London: KPI.

Nakane, C. (1970, 1973), *Japanese Society*, Harmondsworth: Penguin.

National Institute of Employment and Vocational Research (NIEVR) ed. (1984a), *Report of Study Committee Concerning Impact of Microelectronics on Employment*, March, Tokyo.

(1984b), *Problems and Prospects of In-Company Human Resources in a New Era*, November, Tokyo.

(1983), *Summary of Results on the Development of Human Resources in Periods of Technological Innovations and Aging Workforce*, January, Tokyo.

See also *Koyo shokugyo sogo kenkyujo*.

NEDO (1986), *Changing Work Patterns*, Report prepared by the IMS for NEDO in association with the Department of Employment.

Nicholas, I., M. Warner, A. Sorge and G. Hartmann (1983), 'Computerized Machine Tools: Manpower Training and Skill Polarization' in G. Winch ed. *Information Technology in Manufacturing Processes*, London: Rossendale.

Nihon Ginko (1988), *Kokusai hikaku tokei* (Comparative Economic and Financial Statistics), Tokyo.

Nihon kogyo shinbun sha ed. (1986), *Nihon kogyo nenkan* (Japan Industrial Annual), Tokyo.

Nihon kosaku kikai kogyokai ed. (1988), 'Showa 62 nen NC kosaku kikai seisan jisseki to chosa' (1987 CNC Machine Tool Production Results Survey), Tokyo.

(1982), *Haha naru kikai: 30 nen no ayumi* (Mother Machine: 30-Year Path), Tokyo.

See also Japan Machine Tool Builders' Association (JMTBA).

Nihon rodo kyokai ed. (1984), *Maikuroerekutoronikusu kiki no donyu to rodo kumiai no taio* (Introduction of Microelectronics Machines and the Response of Labour Unions), Tokyo.

Nenkan nihon no roshi kankei (Annual of Japanese Industrial Relations), various years.

See also Japan Institute of Labor (JIL)

Nihon seisansei honbu (1987), *Katsuyo rodo tokei* (Practical Labour Statistics), Tokyo.

ed. (1986), *Jinji romu bumon no yakuwari to roshi kankei no shorai* (The Role of the Personnel Department and the Future of Industrial Relations), Tokyo.

ed. (1982), *Saisentan gijutsu to soshiki, ningen* (Advanced Technology and Organizations, People), Tokyo.

See also Japan Productivity Centre (JPC)

Nikkeiren ed. (1985), *Chusho kigyo no seisansei, chingin, roshikankei* (Productivity, Wages and Industrial Relations in Small and Medium-Sized Enterprises), Tokyo.

Nishiguchi, T. (1989), 'Japanese Subcontracting: Evolution Towards Flexibility', D.Phil. dissertation, Oxford University.

Nishiyama, T. (1984), 'The Structure of Managerial Control: Who Owns and Controls Japanese Businesses?' in Sato and Hoshino ed. *The Anatomy of Japanese Business*, New York: Sharpe.

(1975) *Gendai kigyo no shihai kozo* (The Structure of Domination of Modern Corporations), Tokyo: Yuhikaku.

Niyusu daijesto sha ed. (1984), *Kosaku kikai fukutokuhon* (Machine Tool Reader), 7th edition, Nagoya.

(1974), *Kosaku kikai: 10 nen no kiseki* (Machine tools: the 10-Year Path), Nagoya.

Noble, D. (1984), *Forces of Production*, New York: Alfred Knopf.

(1979), 'Social Choice in Machine Design' in A. Zimbalist ed. *Case Studies on the Labour Process*, New York: Monthly Review Press.

Northcott, J., and P. Rogers (1982), *Microelectronics in Industry: What's Happening in Britain*, London: PSI.

Northcott, J., P. Rogers and A. Zeitlinger (1984), *Microelectronics in British Industry: the Pattern of Change*, London, PSI.

Northcott, J., P. Rogers, M. Fogarty and M. Trevor (1985), *Chips and Jobs: Acceptance of New Technology at Work*, London: PSI.

Nyman, S., and A. Silberston (1978), 'The Ownership and Control of Industry', *Oxford Economic Papers*, vol. 30 no. 1, March.

OECD (1977), *The Development of Industrial Relations Systems: Some Implications of the Japanese Experience*, Paris.

Okuda, K. (1982), *Nihongata keiei o ikasu* (Using Japanese-Style Management), Tokyo: Nihon seisansei honbu.

Okumura, H. (1988), *Nihon no kabushiki shijo* (The Japanese Stock Market), Tokyo: Daiyamondo sha.

(1984), *Hojin shihonshugi* (Corporate Capitalism), Tokyo: Ochanomizu shobo.

Ouchi, W. (1981), *Theory Z*, New York: Addison Wesley.

Palmer, G. (1983), *British Industrial Relations*, London: Allen and Unwin.

Parsons, T., and E. Shils (1951), *Toward a General Theory of Action*, Cambridge MA: Harvard University Press.

Pascale, R. and A. Athos (1981), *The Art of Japanese Management*, New York: Simon and Schuster.

Penn, R. (1982), 'Skilled Manual Workers in the Labour Process, 1856–1964' in Wood ed. *The Degradation of Work?*, London: Hutchinson.

Plath, D. ed. (1983), *Work and Lifecourse in Japan*, Albany NY: SUNY.

Pollert, A. (1987), 'The Flexible Firm: A Model in Search of Reality (or a Policy in Search of Practice)?', *Warwick Papers on Industrial Relations*, #19.

Prais, S. (1987), 'Educating for Productivity: Comparisons of Japanese and English Schooling and Vocational Preparation' in *National Institute Economic Review*, London: NIERS.

Pugh, D. (1973, 1984), 'The Measurement of Organization Structures: Does Context Determine Form?' in D. Pugh ed. *Organizational Theory: Selected Readings*, Harmondsworth: Penguin.

Purcell, J., and K. Sisson (1983), 'Strategies and Practice in the Management of Industrial Relations' in G. Bain ed. *Industrial Relations in Britain*, Oxford: Basil Blackwell.

Richardson, R., and A. Nejad (1986), 'Employee Share Ownership Schemes in the UK – an Evaluation', *British Journal of Industrial Relations*, Vol. 24, No. 2, July.

Robins, K., and F. Webster (1982), 'New Technology: a Survey of Trade Union Response in Britain', *Industrial Relations Journal*, Vol. 13.

Rodosho (1987), *Nihonteki koyo kanko no henka to tenbo* (Changes and Prospects in Japanese-style Employment Practices), Tokyo: Okurasho insatu kyoku.

(1986), *2000 nen no rodo* (Labour in the Year 2000), Tokyo: Okurasho insatu kyoku.

(1986, 1985a), *Chingin kozo kihon tokei chosa* (Basic Statistical Survey on Wage Structures), Tokyo: Okurasho insatu kyoku.

(1986, 1988), *Rodo tokei yoran* (Handbook of labour Statistics), Tokyo: Okurasho insatu kyoku.

(1985b), *Nihon no roshi komiyunikeshion no genjo* (Present Situation of Japanese Labour-Management Communication), Tokyo: Okurasho insatu kyoku.
(various years), *Rodo hakusho* (White Paper on Labour), Tokyo: Nihon rodo koyokai.
Rodosho shokugyo noryoku kaihatsu kyoku ed. (1987), Kokyo shokugyo kunren no genjo (Present State of Public Vocational Training), *Shokugyo noryoku kaihatsu janaru* (Journal of Vocational Ability Development), Vol. 29, No. 7, Tokyo.
Rose, M., and B. Jones (1985), 'Managerial Strategy and Trade Union Responses in Work Reorganization Schemes at Establishment Level' in Knights *et al.* ed. *Job Redesign: Critical Perspectives on the Labour Process*, Aldershot: Gower.
Rueschemeyer, D. (1986), *Power and the Division of Labour*, Cambridge: Polity Press.
Sabel, C. (1982), *Work and Politics: the Division of Labour in Industry*, Cambridge: Cambridge University Press.
Sako, M. (1988), 'Neither Markets Nor Hierarchies: a Comparative Study of Informal Networks in the PCB Industry', paper presented to Conference on Comparing Capitalist Economies: Variations in the Governance of Sectors, Wisconsin.
Schonberger, R. (1982), *Japanese Manufacturing Techniques: Nine Hidden Lessons in Simplicity*, New York: Free Press.
Scott, J., and C. Griff (1984), *The British Corporate Network, 1904–1976*, Cambridge: Polity Press.
Shimada, H. (1983), 'Japanese Industrial Relations – A New General Model?' in Shirai, T. ed. *Contemporary Industrial Relations in Japan*, Madison WI: University of Wisconsin Press.
Shimada, H., and J. MacDuffie (1986), 'Industrial Relations and Humanware', MIT, Sloan School of Management.
Shirai, T. (1986), *Roshi kankei ron* (Theory of Industrial Relations), Tokyo: Nihon rodo kyokai.
ed. (1983), *Contemporary Industrial Relations in Japan*, Madison WI: University of Wisconsin Press.
Shokygyo kunren kenkyu senta ed. (1983), *Maikuroerekutoronikusu jidai no jinzai kaihatsu* (Development of Human Resources in the Microelectronics Age), Tokyo: Okurasho.
Sisson, K. (1983), 'Employers' Associations' in G. Bain ed. *Industrial Relations in Britain*, Oxford: Basil Blackwell.
Sisson, K., and W. Brown (1983), 'Industrial Relations in the Private Sector: Donovan Re-visited' in G. Bain, ed. *Industrial Relations in Britain*, Oxford: Basil Blackwell.
Smart, B. (1985), *Michel Foucault*, Sussex: Ellis Horwood.
Smith, G. (1986), 'Profit Sharing and Employee Share Ownership in Britain', *Employment Gazette*, September.
Sorge, A. and M. Warner (1986), *Comparative Factory Organization: and Anglo-German Comparison of Management and Manpower in Manufacturing*, Hampshire: Gower.

Sorge, A. Hartmann, G. and I. Nicholas (1983), *Microelectronics and Manpower in Manufacturing*, Hampshire: Gower.
Stock Exchange (1987), 'The Stock Exchange Survey of Share Ownership Supplement', London.
Storey, J. (1980), *The Challenge to Management*, London: Kogan Page.
Sumiya, M. (1986), 'Kigyobetsu kumiai no shakai keizai kitei' (The Socioeconomic Base of Enterprise Unions), in Nihon rodo kyokai ed. *Nenpo nihon no roshi kankei* (Annual of Japanese Industrial Relations), Tokyo: Nihon rodo kyokai.
ed. (1985), *Gijutsu kakushin to roshi kankei* (Technological Innovation and Industrial Relations), Tokyo: Nihon rodo kyokai.
Taniguchi, T. (1987), 'Overview of the Singapore Productivity Development Project', paper presented to 1987 Asian Regional Conference on Industrial Relations, Tokyo.
Taylor, J. (1983), *Shadows of the Rising Sun: a Critical View of the Japanese Economic Miracle*, Tokyo: Tuttle.
Thomas, R. (1988), 'Technological Choice: Obstacles and Opportunities for Union Involvement in New Technology Design', MIT, Sloan School of Management Working Paper, 1987–88.
Thompson, E.P. (1963, 1968), *The Making of the English Working Class*, Harmondsworth: Penguin.
Thurley, K. and S. Wood eds. (1983), *Industrial Relations and Management Strategy*, Cambridge: Cambridge University Press.
Tokunaga, S. (1983), 'A Marxist Interpretation of Japanese Industrial Relations With Special Reference to Large Private Enterprises' in Shirai, T. ed. *Contemporary Industrial Relations in Japan*, Madison WI: University of Wisconsin Press.
Tokunaga, S., and J. Bergmann (1984), *Industrial Relations in Transition: the Case of Japan and the Federal Republic of Germany*, Tokyo: University of Tokyo Press.
Toritsu rodo kenkyujo (1983), *Chusho kigyo bunya ni okeru sangyo betsu rodo kumiai* (Industrial Unions in the Small and Medium-sized Enterprise Sector), Tokyo.
Totsuka, H., T. Hyodo, K. Kikuchi and M. Ishida (1987), *Gendai Igirisu no Roshi Kankei* (Industrial Relations in Modern Britain) Vol. 1, 2, Tokyo: University of Tokyo Press.
Toyo keizai shinbunsha (1987), *Kaisha shikiho* (Company Information Quarterly), Tokyo.
Trades Union Congress (TUC) (1979a), *Employment and Technology*, London.
(1979b), *Trades Union Congress Economic Review 1979*, London.
Trent, E. (1984), *Metal Cutting*, 2nd edition, London: Butterworths.
Tsusansho ed. (1984), *FA ga kojo o do kaeruka* (How Does Factory Automation Change the Factory?), Tokyo: Nihon noritsu kyokai.
Turner, H. (1962), *Trade Union Growth, Structure and Policy*, London: Allen and Unwin.
Ujihara, S., M. Tsuda, M. Takanashi, K. Koshiro and K. Yamaguchi (1986), '21 seiki o tenbo toshita roshi kankei' (Industrial Relations in View of 21st Century Prospects) in *Kokiro kenkyu*, Tokyo: Kokiro senta.

Urabe, K. (1984), *Nihonteki keiei wa shinka suru* (Japanese-style Management Evolves), Tokyo: Chuo keizaisha.
Vogel, E. (1985), *Comeback*, New York: Simon and Schuster.
ed. (1975), *Modern Japanese Organization and Decision Making*, California: University of California Press. Tuttle edition, 1979.
Warner, M. (1984), 'New Technology, Work Organizations and Industrial Relations' *Omega International Journal of Management Science*, Vol. 12, No. 3.
Watanabe, Y. (1983, 1984), 'Shitauke kigyo no kyoso to sonritsu keitai' (Subcontractor Competition and Patterns of Being), *Mita gakkai zasshi*, Vol, 76, No. 2; Vol. 76, No. 5; Vol. 77, No. 3.
Webb, S., and B. Webb (1902), *Industrial Democracy*, London: Longmans.
Weber, M. (1968). *Economy and Society*, Vols. 1–3, New York: Bedminster.
(1930, 1976), *The Protestant Ethic and the Spirit of Capitalism*, London: George Allen and Unwin.
Werskey, G. (1987), 'Training for Innovation: How Japanese Electronics Companies Develop Their Elite Engineers', London: GEC.
White, M. (1981), *Payments Systems in Britain*, Aldershot: Gower.
(1979), *The Hidden Meaning of Pay Conflict*, London: Macmillan.
White, M., and M. Trevor (1983) *Under Japanese Management*, London: PSI.
Whittaker, D. (1988), 'New Technology and Employment Relations: CNC in Japanese and British Factories', Ph.D. dissertation, Imperial College, London University.
(1989), 'The End of Japanese-style Employment?' US–Japan Relations Program Occasional Papers, Harvard University.
Wiener, M. (1981), *English Culture and the Decline of the Industrial Spirit, 1850–1980*, Cambridge University Press.
Wilkinson, B. (1983), *The Shopfloor Politics of New Technology*, London: Heinemann.
Williamson, O. (1975), *Markets and Hierarchies: Analysis and Antitrust Implications; a Study in the Economics of Internal Organization*, New York: Free Press.
Williamson, O., and W. Ouchi (1983), 'The Markets and Hierarchies Programme of Research: Origins, Implications and Prospects' in Francis *et al.* eds. *Power, Efficiency and Institutions*, London: Heinemann.
Willman, P. (1983), 'The Organizational Failures Framework and Industrial Sociology' in Francis *et al.* eds. *Power, Efficiency and Institutions*, London: Heinemann.
Winch, G. ed. (1983), *Information Technology in Manufacturing Processes*, London: Rossendale.
Winchester, D. (1983), 'Industrial Relations in the Public Sector' in G. Bain ed., *Industrial Relations in Britain*, Oxford: Basil Blackwell.
Wood, S. (1982). *The Degradation of Work?* London: Hutchinson.
Yahata, S. (1985), 'FA ka to shokumu naiyo no henka' (Factory Automation and Changes in Job Contents) in *Kikan Rodoho* (Labour Law Quarterly), No. 137.
Yakabe, K. (1984), 'Shingijutsu jidai ni okeru roshi kankei no tenbo' (Outlook for Industrial Relations in an Age of New Technology), in *Saisentan jigutsu to*

Soshiki, Ningen (Leading Edge Technology and Organizations, People), Tokyo; Nihon Seisansei Honbu.

Yamaguchi, K. (1983), 'The Public Sector: Civil Servants' in T. Shirai ed. *Contemporary Industrial Relations in Japan*, Madison WI: University of Wisconsin Press.

Zimbalist, A. ed. (1979), *Case Studies in the Labour Process*, New York: Monthly Review Press.

Index

Abegglen, J. and G. Stalk, 4, 99
age
  -earnings 22–3, 70, 71, 75
  in selection of CNC operators 125–7
  of operators 106–7, 126, 127, 157, 158
  of recruits 56, 58, 65, 95, 97, 107, 157, 158
Amalgamated Engineering Union (AEU) 13, 77, 85, 86, 101, 148, 159
Amalgamated Union of Engineering Workers (AUEW) *see* Amalgamated Engineering Union
apprentices 6–7, 55, 58, 65, 74, 116, 117, 119, 131, 171
apprenticeships 11, 17, 56, 60, 64, 117–18, 119, 120, 124, 128, 134
Arthurs, A. 51
assembly 62, 63, 118, 143
Association of Professional, Executive, Clerical and Computer Staff (APEX) 82, 84
Association of Scientific, Technical and Managerial Staffs (ASTMS) 84

Babbage, C. 154
Banking Insurance and Finance Union (BIFU) 37
batch size 2, 14–16, 17, 19, 102, 111, 125, 135, 136, 141, 143, 144, 145, 153–4, 159
Batstone, E. 2
Berger, S. and M. Piore 36
Braverman, H. 1–2, 14, 101, 144, 161
British Workplace Industrial Relations (BWIR) survey 25, 26, 28, 29, 52
Brown, Wildred 38
Brown, William 24, 33–4, 50–1, 52

capitalism 1, 2, 39, 42–3
Child, J. 4
Clark, R. 25, 36, 41, 42
Clearing Banks Union 37
Cole, R. 49
collective bargaining, negotiations 24, 33, 46, 69, 71, 81, 83, 84, 85, 87, 90
communications 10, 26, 32, 50, 53, 63, 83, 86, 87, 88, 91, 96, 97, 151
communist party 82, 87
community
  corporate 57, 65, 78
  local 108
  of interests, 'fate' 31, 35, 42, 43, 47, 162
company newspapers, magazines (*shanaiho*) 66, 86
compliance 8, 12, 45, 47, 48, 49
computer aided design/computer aided manufacture (CAD/CAM) 93, 94, 104, 105, 106, 164
computer numerical control (CNC) machine tools 1–3, 4, 5, 12, 99–100, 156–7, 164
  'allergy' 125, 126
  automation 17, 104–8
  development 100–1
  'language speakers' (*NC gengozoku*) 157–8
  purchase 92–3, 101–4, 106–7, 130, 131, 132
Confederation of British Industry (CBI) 41
consultation 7, 26, 28, 81, 82, 83, 92–4, 96, 104
contractual relations 6, 8, 9, 10, 11, 32, 33, 42, 53, 74
craft
  approach to CNC 5, 156, 160, 161, 162
  organization 10, 17, 162
  training 64, 114, 116–19, 124, 136, *see also* apprenticeships

Craft (*cont.*)
workers, craftsmen 1, 5, 57, 60, 61, 77, 117, 118, 119, 124, 125, 140, 144, 148, 154, 159
culture 29–30, 32, 34, 44, 47
organization 113

Daniel, W. 28
deburring 13, 121, 140, 143
demarcation 15, 140, 148
Denkiroren (Japanese Federation of Electrical Machine Workers Unions) 27
Department of Employment 21, 22, 23
design, draughting 55, 61, 102, 169
direct numerical control (DNC) 104, 105, 106, 150, 164
directors 42–3, 54, 64, 65, 89, 91, 167, 168, 169, 170
Dobson, C. 33
Dodgson, M. 13, 17
Domei (Japanese Confederation of Labour) 28, 86
Donovan Commission 24
Dore, R. 9–10, 25, 33, 36, 41, 115, 116
Doro (National Railway Motive Power Union) 34
dualism 35, 36

Economic Planning Agency (Keizai kikakucho) 5, 20–1, 45, 48
education 54, 55, 60–1, 115–16, 127, 136, 163
Edwards, P. and H. Scullion 38
Electrical, Electronic, Telecommunication and Plumbing Union (EETPU) 52
Elliot, R. and J. Fallick 37
employee participation, involvement 28, 39, 109, 110, 112
employment
'fixed' 21–2, 48, 59
length of 20–2, 57–9
lifetime 2, 5, 20–1, 31, 37, 59
multitrack 49, 50
non-regular 48, 166, 167, 168, 169, 170, 171
employment relations 5–6
changes in 48–52, 90–1
evolution of 31–4
sector differences 34–5, 36–7
see also organization-oriented employment relations
Engineering Industry Training Board (EITB) 63
engineers 54, 60, 64, 66, 132, 148, 157
production 14, 93, 104, 130, 145, 146, 147, 148, 150, 161–2
Etzioni, A. 8
evaluations 7, 64, 67–8, 74, 75–6, 95, 97

factory size 17–18
CNC use 2, 14–15, 100, 102, 103, 104, 145–7, 153–5, 157–8
employment relations 15, 18, 21, 24, 35, 36, 39, 56, 65, 89, 91, 97
wages 35, 38, 71–2, 76, 78–9
female workers 21, 22, 23, 27, 38, 48, 54, 65, 143, 148, 166
fitting 61, 73
flexible manufacturing cell (FMC) 105, 107
flexible manufacturing system (FMS) 105, 106, 107, 164
flexibility 2, 4, 7, 11, 38, 51, 52, 62, 90, 109, 138, 139, 142, 155
pay and 49, 74, 79, 95, 97
foreman *see* supervision
Foucault, M. 45
Fox, A. 9, 13–14, 20, 32, 33, 44, 90
Francis, A. 12, 39, 159
Friedman, A. 14, 38

General Municipal, Boiler Makers and Allied Trades Union (GMBATU) 82, 83
Gennard, J. 38
gino kentei (skill testing) 120
goal convergence, congruence 8, 46–7, 109–10, 113
Gordon, A. 31
Grayson, D. 51
groupism (*shudanshugi*) 3, 30, 34

Hanami, T. 47
harmonization 7, 23, 51, 53, 74–5, 76–7, 79, 91, 169
Hattori, T. 161
hierarchical observation 45, 46
Hobsbawm, E. 17, 44
honko (regular employees) 35–6
human resource management 3, 5, 52, 53, 64, 65, 66, 81, 91
Hyman, R. 45
hypotheses 3, 10–17, 114, 133–7, 151–5, 158–60, 161–3

Inagami, T. 26, 34, 35, 50
individualism 30, 32, 34
industrial action 28, 34, 36, 44, 85, 87–8, 89–90, 91, 93, 140–1
industrial relations

contours of 7, 13, 17, 24–6, 52, 53
organization of in factories 25, 80–5, 87, 89
recent changes 49–50, 52, 90–1
industrialization 9–10, 29, 33
Ingham, G. 38–9, 46
innovation
attitudes towards 3–4, 27–9, 51, 106–8, 112–13, 162
CNC, etc. 5, 99–108
linkage 4, 104–6, 111–13
'soft' 4, 108–11
see also new techonology, craft approach, technical approach
inspection 13, 118, 140, 142, 143, 148, 149
'intellectualization' of blue collar work 156
Ishikawa, T. 120
Itami, H. 26, 49

Japan Institute of Labour (Nihon rodo kyokai) 29
Japan Machine Tool Builders Association (Nihon kosaku kikai kogyokai) 101
Japan Productivity Centre (Nihon seisansei honbu) 25, 43, 61, 64
job definitions, boundaries, 7, 13–14, 62, 77, 90, 95, 97, 109, 138, 142–4, 155 *see also* demarcation, task range
Jones, B. 2, 13, 62
just in time (JIT) 4, 99, 105, 108, 111

Kaplinsky, R. 4
Keizai doyukai (Japan Committee for Economic Development) 50
Keizai kikakucho *see* Economic Planning Agency
Kelley, M. and H. Brooks 17
Koike, K. 11, 20, 22, 24, 26, 119
Koseki, T. 157–58, 160
Koshiro, K. 34, 50
Koyo shokugyo sogo kenkyujo *see* National Institute for Employment and Vocational Research
Kuwahara, Y. 42, 51

labour costs 37, 49, 70, 75
labour markets
external 7, 11, 45, 48, 51, 78, 130
intermediate 49
internal, internal careers 4, 6, 7, 11, 49, 50–1, 60–1, 78, 91, 119–20
Lee, D. 17
Leeson, R. 33
Levine, S. 20

Littler, C. 2, 3, 33
Lloyd, P. and R. Blackwell 36

Macfarlane, A. 32
McGregor, D. 8
machi koba ('street corner' factories) 18, 106, 157
machine tools (especially conventional) 1, 12, 17, 19, 100–1, 103, 111, 112, 114, 130, 155 *see also* computer numerical control (CNC) machine tools
management
involvement 161
strategies 2, 3, 8, 15, 17, 98, 103, 159, 160, 161, 169
managers
backgrounds of 60, 91–2, 128, 137
views of CNC 101–2, 106–8, 111, 112, 116–17, 121, 122–5, 126–7, 128, 137, 143–4, 145–7, 156–7, 159, 160
manual data input (MDI) 100, 102, 144, 164
market-oriented employment relations (MER) *see* organization oriented employment relations (OER)
markets and hierarchies 8
Martin, R. 32, 44
Mihara, Y. 120
Millward, N. and M. Stevens 28, 38
Ministry of International Trade and Industry (Tsusansho) 12, 159
Ministry of Labour (Rodosho) 28, 49
mobility
external 30–1
horizontal 4, 61–2, 119–20, 151, 155
vertical 4, 8, 60–1, 62, 119–20, 151
see also turnover
module production 105, 108–9, 111
morale 11, 49, 77, 113, 146, 162
Mori, K. 106
Mouer, R. and Y. Sugimoto 30

National Institute of Employment and Vocational Research (koyo shokugyo sogo kenkyujo) 14, 15, 159
New Earnings Survey 24, 38
new technology 1–2, 3–4, 55, 77, 90, 99, 104, 120, 161
employment relations 1–3, 26, 161–3
industrial relations 26–9, 92–4
unemployment 29, 92
Nihon kosaku kikai kogyokai *see* Japan Machine Tool Builders Association
Nihon rodo kyokai *see* Japan Institute of Labour

Index

Nihon seisansei honbu *see* Japan Productivity Centre
nihonjinron (theories of Japanese character) 30
Nikkagiren (Japan Union of Scientists and Engineers) 109
Nikkeiren (Japan Federation of Employers' Associations) 37, 43
Nishiguchi, T. 35
Nishiyama, T. 42
Noble, D. 1, 12, 100
norms, 8, 11, 26, 45, 47, 48, 53, 56, 57, 58, 59, 62, 78
numerical control (NC) 1, 99–100, 101, 164
 see also computer numerical control (CNC) machine tools
Nyman, S. and A. Silberston 39, 41

Odaka, K. 30
Okumura, H. 41, 42, 43
operating 13, 15, 16, 138–41, 145, 152
 parallel, multimachine 142, 143–44, 157
 unmanned 5, 106–8, 143, 157, 163
operators
 ages of *see* age
 backgrounds of 115–16, 117–19, 120–1, 122–5
 friction with programmers 13, 147, 150–51
 trends 102, 109, 117, 124, 125
organization-oriented employment relations (OER) and market-oriented employment relations (MER)
 British and Japanese factories 9, 20, 39, 48, 50, 96–8
 concept 6–9
 eighteen factories 94–8
 employment 6, 7, 53
 industrial relations 7, 26, 53
 pay 6–7, 22–3, 53, 78
 task range and 13–14, 151, 153, 155
 training and 10–12, 114, 133–7
orientation courses 54–5, 57, 65, 95
ownership 6, 35, 39–44 *see also* shares, shareholding

Palmer, G. 38
Parsons, T. 8
Pascale, R. and A. Athos 46
Patternmakers, 82, 83
payment systems, wages
 age and *see* age
 allowances 69, 72, 77
 bonus, productivity scheme 69–70, 72, 74, 76
 British 37, 73–77
 CNC and 11, 13, 73, 77–9
 grades 71, 73, 74, 68–9, 75–6
 holiday pay 75
 Japanese 2, 5, 22–3, 37, 49, 66–72
 job-based 7, 11, 13, 74, 78
 'living' wages 33, 49, 67, 69, 71
 nenko see payment, Japanese
 person-related 7, 11, 13, 67–71, 76, 78
 piecework 38, 66, 73, 87
 public sector 34, 36–7
 sick pay 75, 77
personnel function, management 7, 53, 54, 58, 61, 62–5, 80, 83, 91, 108, 110, 113, 120, 121, 165, 168, 170
power 8, 13, 43, 44–8, 86, 93
Prais, S., 115
preventative maintenance 110, 121, 142, 143
production engineering department 61, 73, 102, 104, 147, 148, 150, 161
products 4, 17–18, 123, 153, 166, 167, 168, 169, 170
programmers 13, 92, 130, 138, 148–151, 157, 159
programming 1, 2, 12, 13, 14, 15, 16, 100, 102, 126, 138, 140, 144–8, 152, 157, 159
promotion *see* mobility

quality control (QC) 32, 86, 88, 99, 108–10, 161

recruiting 7, 37, 53, 54–7, 91, 94, 95, 97, 115, 119, 134
redundancies 28, 57, 58, 62, 90
Rengo (Japanese Private Sector Trade Union Confederation) 28, 89
resources 15, 24, 26, 36, 44–8, 85–6, 91, 96, 135, 157
retirement 29, 37, 57, 58, 67, 70, 72, 166, 167
Rodosho *see* Ministry of Labour
Rose, M. 2
rotation *see* mobility
Reuschmeyer, D. 44

Sako, M. 35, 115, 116
Salaman, G. 2
Scott, J. and C. Griff 41, 42
seniority system *see* payment, nenko
setters 73, 109, 139, 140, 141
setting 13, 14, 15, 16, 117, 119, 145, 152, 159
shagaiko ('outside' workers) 35
shares, shareholding 8, 42–3, 89, 167, 168,

169, 170
by employees 51, 167, 168
shifts 107, 132, 163
Shirai, T. 26, 35
shop stewards *see* unions
shukko ('loan' workers) 49, 55, 168
single status *see* harmonization
Sisson, K. 24
skill centre, polytechnic 122, 130–1
skills
concept 16–17
craft 101, 128, 156, 158, 161, 163
technical 111–12, 115, 156, 158, 161
slingers 13, 140, 141
slogans and mottoes 45, 47, 50, 111–13, 156, 162
small group activities 108–10 *see also* quality control
Smith, G. 51
socialization 15, 53, 65, 115, 163
Sohyo (General Council of Trade Unions of Japan) 28, 88
Sorge, A. et. al. 2, 14–15, 153, 163
strikes *see* industrial action
subcontracting 35, 90, 166, 168, 169
suggestion schemes 109, 110
Sumiya, M. 31, 35
supervision 68, 69, 71, 76, 89, 92, 102, 104, 115, 117, 119, 130, 132, 146, 147, 150
intensity of 46, 62, 117, 121, 127–8, 136
personnel department and 63, 64, 108

Taniguchi, T. 43
task range 15–16, 62, 138, 139–48, 151–3 159
employment relations and 13–14, 17, 153–5
Taylor, J. 30
Taylorism 3, 14, 138, 154
Technical, Administrative and Supervisory Section (TASS) 13, 82, 84, 93, 148, 159
technical approach to CNC 5, 156, 160, 161, 162, 163
Theories X, Y 8
three pillars 2, 5, 6, 20, 26, 48, 80
Tokunaga, S. 44
tool-room 61
Toritsu rodo kenkyujo (Tokyo Metropolitan Labour Research Institute) 24, 25
Totsuka, H. *et al.* 161
Trades Union Congress (TUC) 26–7, 52, 86
training
employment relations and 3, 10–12, 135–7
external 11, 122, 128–31
factory size 122–5, 135
off-job 12, 54, 55–6, 131–2
on-job (OJT) 114, 119–21, 132
orientation 7, 54–5, 56, 57, 65
personnel policies and 62–4
unions and 27, 28, 86, 91
Transport and General Workers Union (TGWU) 85
Tsusansho *see* Ministry of International Trade and Industry
turnover 21, 30, 31, 38, 54, 58, 65, 132

Ujihara, S. 38
Unions
attitudes towards 88, 90–1
committees 25, 26, 81, 82, 83, 84, 85, 87, 90
enterprise 2, 5, 6, 24, 25, 31, 49, 80, 86
federations, confederations 24–5, 26, 27, 28, 86, 88, 89
general meetings 81–2, 83
officials 52, 85, 86, 87, 91–2
organization 25, 80–5
representatives, stewards 25, 26, 28, 33, 52, 80, 81, 83, 84, 86, 91, 92, 93
resources 25–6, 85–6
second 25, 82
surveys 46, 86, 88
Us-Them 9, 34, 76, 150, 162

wages *see* payment systems, wages
Warner, M. 99, 163, *see also* Sorge et al.
Webb, S. and B. 32
Weber, M. 44
Werskey, G. 47
West Germany 14, 163
White, M. 38
'white collarization' of blue collar worker 112
Williamson, Oliver 8
Willis, N. 52
Winchester, D. 36
Wood, S. 14
works committees, councils 83, 84, 85

Yamaguchi, K. 34

Zenkoku kinzoku (National Federation of Metal and Engineering Workers) 25, 86